D0024646

The Cultural Geography of the United States

A Revised Edition

Wilbur Zelinsky

Professor Emeritus of Geography
The Pennsylvania State University

PRENTICE HALL, Englewood Cliffs, N. J.

Library of Congress of Cataloging-in-Publication Data

ZELINSKY, WILBUR,
    The cultural geography of the United States / by Wilbur Zelinsky.
— A rev. ed.

        p.   cm.
    Includes bibliographical references and index.
    ISBN 0-13-194424-X
        1. United States—Civilization.   2. Human geography—United
States.   3. National characteristics, American.   I. Title.
    E169.1.Z38 1992   91-34991
    973—dc20   CIP

Editorial/Production Supervision and Interior Design: Judi Wisotsky
Acquisition Editor: Ray Henderson
Copy Editor: James Tully
Cover design: Lundgren Graphics
Prepress Buyer: Paula Massenaro
Manufacturing Buyer: Lori Bulwin

 © 1992, 1973 by Prentice-Hall, Inc.
A Simon & Schuster Company
Englewood Cliffs, New Jersey 07632

Printed in the United States of America

10 9 8 7 6 5 4 3 2 1

ISBN 0-13-194424-X

Prentice-Hall International (UK) Limited, *London*
Prentice-Hall of Australia Pty. Limited, *Sydney*
Prentice-Hall Canada Inc., *Toronto*
Prentice-Hall Hispanoamericana, S.A., *Mexico*
Prentice-Hall of India Private Limited, *New Delhi*
Prentice-Hall of Japan, Inc., *Tokyo*
Simon & Schuster Asia Pte. Ltd., *Singapore*
Editora Prentice-Hall do Brasil, Ltda., *Rio de Janeiro*

Contents

Tables *vii*
Figures *vii*
Preface *ix*

PART **1** *background*

CHAPTER ONE
Origins 3

Why the United States? 3
Basic Processes of American Cultural Development 5
Parallels with Other Parts of the World 9
The Cultural Legacy of the Old World 10
The Aboriginal Contribution 14
The African Contribution 18
The Continental European Contribution 20
Significance of Sequent Waves of Immigration 22
Territorial Patterns among Ethnic Groups 28
The Turner Thesis 33
Why Did a National Culture Come to Be? 35

CHAPTER TWO
Identity 36

On Capturing the American National Character 36
The Theme of Individualism in American Culture 41
Individualism Manifested in the Cultural Landscape 44
Political Patterns 44   Settlement Pattern 47

iii

Economic Patterns *50*   Religion and Education *51*
Miscellaneous Symptoms *52*
The Theme of Mobility and Change 53
The Theme of a Mechanistic World Vision 59
The Theme of Messianic Perfectionism 61

PART   *spatial expression*

CHAPTER THREE
Process   *67*

Explaining the Spatial Aspects of Cultural Change 67
Some Basic Cultural Propositions 68
Population Mobility as a Factor in Cultural Change 77
The Diffusion of Innovations 79
Convergence vs. Divergence *85*
The American House 88
Geography of Denominational Membership 94
Other Aspects of American Culture 100
Settlement Features *100*
Cultural Demography and Cultural Economics *103*
Language *105*
Institutions, Social Behavior, and Folklore *107*

CHAPTER FOUR
Structure   *109*

Why Study American Culture Areas? 109
The Classification and Anatomy of Culture Areas 110
The Major Traditional Culture Areas 117
An Historical Note *117*   New England *120*
The South *122*   The Midland *125*   The Middle West *128*
The Problem of the "West" *129*
Lesser-Order Culture Areas *133*
The Voluntary Region 134
Educational Subregions *136*   The Pleasuring Places *136*
Latter-Day Bohemias and Utopias *138*
Postlude 139

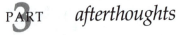

PART   *afterthoughts*

CHAPTER FIVE
America in Flux   *143*

Introduction 143   *The Ambient Scene 144*
Militarism and Violence *146*   Nationalism *149*

Effects of Technological Change in Communications
and Transport *153*
Transnationalization of Culture *156*
Restructuring of Sensibilities and Places *160*
Ethnic Developments *162*   Shifting Settlement Patterns *164*
The New Landscape Awareness *169*
Regional/Ethnic/Historic Revivals *173*
The American Culture Area Revisited *177*
The Question of Convergence *182*   *Envoi* *184*

Selected References   *186*

Index   *213*

List of Tables

1.1 Religious Congregations: United States, 1775–1776   13
1.2 Immigrants to the United States, 1820–1969, by Region or Nation of Origin and Percent of Decadal Total   27
3.1 Regionally Distinctive Patterns of Consumer Expenditures for Selected Goods and Services, 1960–1961   106
5.1 A Typology of Culture Areas   178

List of Figures

1.1 Regionalization of Twelve Major Ethnic Groups in 1960s   30–31
3.1 A Schematic, Three-Dimensional Representation of Cultural Systems   73
3.2 The Movement of Ideas in the Eastern United States   81
3.3 Major Religious Regions   97
3.4 Age-Adjusted Homicide Rate, White Population, Three-Year Average, 1959–1961   104
4.1 The Mormon Culture Region   115
4.2 The Geographic Morphology of the Texas Empire   116
4.3 Culture Areas of the United States   118–119
4.4 The Culture Areas of Texas in the 1960s   125

PREFACE

When the original version of this volume appeared in 1973, I had no idea what sort of reception it would have. My uncertainty derived from a basic consideration, that I had directed it at a readership of one: myself. There was then no other book-length treatment of the cultural geography of my favorite country—nor is there any now other than the present item—and I yearned for something succulent to read. The fact that other readers seemed to enjoy it, that the critical reaction was favorable, that it was inflicted on so many students, and that it somehow kept on selling doggedly year after year in reassuring volume was a pleasant surprise.

After the editor plied me with lunch, drinks, and pitiful entreaties a year ago, I sat down to reread the book from cover to cover, and I discovered, again to my surprise, that it had aged gracefully and had not yet become quaint. After much furrowing of brow, I decided to leave the original four chapters intact instead of attempting a rewrite—certainly not the easiest of decisions. The only modifications in the older material are the correction of several typographical and grammatical errors and the recasting of three brief passages that had become factually flawed.

The principal reason why aficionados of the original edition should read, and perhaps even purchase, this new improved opus is the addition of a meaty fifth chapter entitled "America in Flux." In it I try, as best I can, to describe and interpret the changes in American cultural patterns in the 1970s and 1980s, to exploit the relevant scholarship that has appeared in recent years, and to offer such overall reflections as I was incapable of in the not too distant past. I also urge readers, new and old alike, to read the greatly expanded "Selected References" section with special care. These pages could well comprise the most valuable part of the book. Enjoy!

I would dearly love to hear from readers with their reactions to my efforts, be they good, bad, or indifferent.

Wilbur Zelinsky
University Park, PA

to the land I have loved the most

PART 1 *background*

CHAPTER ONE

Origins

Why the United States?

"What, then, is the American, this new man?"[1] After nearly two centuries, de Crévecoeur's haunting question still cries out as piercingly and urgently as ever. Wherever his home and whatever his calling, the concerned citizen of the late twentieth century must persist in the endless job of interpreting and understanding the most powerful and innovative nation in the world. The literature on the United States—on its land, its people, and their ways and institutions—has grown to mountainous proportions; but new questions arise constantly, and more quickly than the old ones can be answered.

It is to one of these emergent themes that this volume—the first attempt to view the cultural geography of the United States synoptically—is addressed. In what ways has the particular configuration of the American cultural system shaped spatial processes and structures in this nation? How has it molded the visible, and invisible, features of the landscape, or the ways in which the American has evolved his behavior over space and his perceptions of, and interactions with, other people, his habitat, and with ideas both near and far?[2] And how, in turn, have these areal facts modified and channeled the development of the cultural system? In brief, what can a reading of the spatial dimensions of American culture contribute to the unfinished chore of explaining this mighty, dynamic outpost of mankind's future? Obviously, "America watching" is a mandatory activity for those who care about the welfare of the species *Homo*

[1] J. Hector St. John de Crèvecoeur, *Letters from an American Farmer* (London: Thomas Davies, 1782), p. 51.

[2] The concept of culture is defined and discussed in detail in A. L. Kroeber, *The Nature of Culture* (Chicago: University of Chicago Press, 1952). For general introductions to the field of cultural geography and its concepts and methods, consult Philip L. Wagner and Marvin W. Mikesell, eds., *Readings in Cultural Geography*

sapiens or for those who are simply smitten with affection for this lovely and fragile planet. It is also an effective strategy for the scholar intent upon tracking down abstract scientific truth.

Indeed for the student in quest of lawlike statements about cultural geography, the United States is almost too good to be true. He would be hard pressed to identify—or invent—a better country for accumulating facts, for asking basic questions, or for experimenting with new approaches. Given the recent origin of American culture, along with the relative simplicity of its areal expression, America is as nearly ideal a laboratory as could be desired. The cultural landscapes of much of the Old World might be characterized as intricate and ancient, of uncertain origin, revealing multiple time levels and the handiwork of many ethnic groups. In the United States, on the other hand, we deal with what we have come to regard, all too ethnocentrically, as a virtual *tabula rasa* at the outset of European settlement. Actors entering a nearly bare stage play out their roles rapidly, directly, and with few complications of plot. In addition, through pure happenstance, the texture of the "natural" landscape upon which American culture evolved and spread so rapidly tends to be coarser than in most parts of Europe, Asia, or Middle America. Its physiographic, climatic, and biotic provinces are broader and less often punctuated by minor irregularities that might catalyze and shelter local cultural peculiarity. Yet the American cultural scene is by no means a dull, uniform, featureless affair; there is sufficient variety from place to place and through time to entice the student of culture, but not so much as to frustrate or bewilder him. All in all, the accessibility of the American record and its relative transparency as to essential cultural structure and process is an open invitation to scholarly investigation.

A further attraction is the almost embarrassing abundance and variety of raw data, from the beginning to the present. The entire process of European exploration, occupation, and development was richly documented by many literate, alert, and fascinated participants and visitors from earliest days onward; and a great store of words, drawings, pictures, maps, digits, and other durable artifacts have been generated. Thanks to the infrequency of domestic disturbances and large-scale property destruction, most of this material endures. The major problem is selection and digestion.

If scientific expediency and an extraordinary availability of facts (neglected though many of them may be) were not adequate excuses for settling on the United States to demonstrate some basic concerns and devices of cultural geography, we can return to the ultimate argument. By definition, all places are unique; but the United States is unique in

---

(Chicago: University of Chicago Press, 1962); Philip L. Wagner, *The Human Use of the Earth* (New York: The Free Press, 1960); George F. Carter, *Man and the Land; a Cultural Geography*, 2d ed. (New York: Holt, Rinehart & Winston, Inc., 1968); Joseph E. Spencer and William L. Thomas, Jr., *Cultural Geography; an Evolutionary Introduction to Our Humanized Earth* (New York: John Wiley & Sons, Inc., 1969); and Samuel N. Dicken and Forrest R. Pitts, *Introduction to Cultural Geography* (Waltham, Mass.: Xerox College Publishing, 1971).

a manner that transcends the parochial pride or interest of its citizens. Quite clearly the United States is the most powerful nation of the contemporary world in both economic and military terms. This nation also retains much of its traditional strength as a moral force, and has had few recent rivals as a generator of science and technology, or as an exporter of tourists, religious and linguistic influences, and a varied cargo of the popular and fine arts. Thus the cultural personality and behavior of American man is literally of worldwide importance, and quite possibly crucial in deciding the survival or further progress of the human species.

Basic Processes of American Cultural Development

The origin and development of some of the distinguishing features of American culture must be sketched before treating its geography. Basic to this entire discussion is the premise that the United States belongs to the Greater European cultural realm, and that in all essentials our culture is derived from that of Northwest Europe and, most particularly, of Great Britain. This is not to claim that non-British or non-European sources have not contributed to the peculiar American cultural blend, for they obviously have, but simply to assert the powerful genealogical fact that British cultural parentage underlies all else. But if one glances at some of the more extreme forms of American cultural development and then compares them with developments in England past or present, any immediate kinship may be difficult to detect. How, then, did American culture come to be what it is today? The following five processes would seem to have been at work and, in combination, to account for that rapid, massive transformation of Old World elements into the singular American compound.

1. *The importation of selected individuals and, hence, selected cultural traits.* As crucial an item as any in the formation of American culture was the fact that the immigrants who shaped it were not random samples of the Old World population from which they were drawn. Instead, they were selected, consciously or otherwise, for specific attributes; and these attributes, in turn, were probably decisive in imparting to the new cultural system the specific shape and direction it was to follow.

2. *The simple fact of long-distance transfer of people and their cultural freight.* The very act of transporting settlers from Europe (or Africa) across the Atlantic to North America served to initiate social and cultural alterations of some consequence. We have here a "sea change" phenomenon: the forced shipboard association for some weeks or months of persons who had been strangers; the all-too-frequent traumatic impact of the voyage itself; the thrust into unfamiliar physical surroundings; the struggle to establish a viable community during the critical early months in America. Inevitably, the transferred group was noticeably modified by this experience. This is illustrated by solicitous attempts at transplanting certain European groups intact and arresting change in their new setting. In the case of New France and New Spain or some minor groups along the Atlantic Seaboard, the product was

still a colony distinct from the antecedent society.[3] One way of interpreting the founding of a New England is that it was in part an effort to restore the pristine purity of a vanishing, perhaps imaginary ideal England. The result, however, was again a community quite different from the intended model.[4] In brief, then, the abrupt relocation of people across great distances inevitably generates cultural change, however zealously the immigrants may try to prevent such change.

3. *Cultural borrowings from the aboriginal populations.* The adoption of certain cultural items from the Amerindians—or, more precisely, the process of cultural interchange between aborigine and newcomer—was a minor factor, except perhaps locally and temporarily, in the creation of the Euro-American cultural system. Nonetheless, the immigrants did not enter a total human vacuum; and there was some transfer of items, despite the great cultural gap among the peoples involved and, especially, the vast disparity in levels of socioeconomic achievement.

4. *The local evolution of American culture.* With the coming of various groups of selected immigrants to North America, their survival past the first critical episodes of contact with a strange land, and perhaps some dealings with the aboriginal folk, there began in earnest the evolution *in situ* of a recognizably American set of cultures, later to form a single national system, through the operation of at least five processes, as follows:

a. There was selection of cultural traits from among those transmitted from overseas. Obsolete or irrelevant items (for example, most of the feudalistic social and landholding practices) withered and died, but others, previously latent or minor but now of major survival value (for example, woodcraft, hunting and fishing techniques, and systems of slave or contract labor), flourished and developed new variants.

b. In a fashion analagous to the genetic drift observed by biologists, small random cultural mutations occurring at all times to all groups may emerge as events of lasting effect amidst minute groups living under isolated conditions. Thus it is quite possible that in the earlier days of colonization, when settlements were small and scattered, miles were longer, governmental control was quite inefficient, and communications were slow and uncertain, pure chance may have nudged cultural evolution in specific directions in different regions.

c. The interaction between new settler and various phases of the physical habitat almost certainly engendered cultural change. This

---

3 The processes of cultural transfer and transformation are treated in illuminating fashion in George M. Foster, *Culture and Conquest; America's Spanish Heritage*, Viking Fund Publications in Anthropology No. 27 (New York: Wenner-Gren Foundation for Anthropological Research, 1960).

4 The phenomenon is well documented in an exemplary case study: Sumner Chilton Powell, *Puritan Village; the Formation of a New England Town* (Middletown, Conn.: Wesleyan University Press, 1963). A similar story of the difficulties of transplanting the traditional social order and mode of landholding is detailed in Richard Colebrook Harris, *The Seigneurial System in Early Canada; A Geographical Study* (Madison and Quebec: University of Wisconsin Press and Les Presses de l'Université Laval, 1966).

came about either more or less directly through the settler's physiological or psychological responses to novel environmental stimuli or, indirectly, as a byproduct of an economy that, in turn, reflects physical circumstances. This description is not intended to revive the hoary controversy over environmental determinism. Indeed any single-factor, unidirectional determinism must be rejected in the quest for geographic explanation. What must be considered instead are complex *interacting* entities, in this instance the human ecosystem, in which man is assumed to be part of an intricate biophysical system. Within such a set of relationships, cause and effect are inextricably enmeshed so that a given phenomenon is seldom unequivocally one or the other. It would be just as unrealistic to expect the land surface, soils, biota, waters, or microclimates of the country to be unaffected by modern American man as to declare that his pattern of bodily and mental activity and, thus ultimately, his cultural configuration were untouched by the nature of the land.[5]

---

5 But, after all these disclaimers, there still remains the stubborn likelihood that the specific size, location, and physical characteristics of the American territory were relevant factors in the shaping of American culture; or, to be more realistic, the ways particular people at particular times and places perceived and interacted with these factors did somehow influence the course of cultural development. Unfortunately for scientific purposes, the United States was formed only once; and we can never know with any assurance what might have happened if all other variables were held constant and the same batch of settlers had elected Brazil, Kamchatka, or South Africa. But suppose that instead of a roughly rectangular tract of 3,000,000 square miles some 3,000 miles from the homeland with a very large package of known and available physical resources, those very same British colonizers had picked a small, bleak, roundish, sub-Arctic island only some 800 miles from their ports of embarkation, as did the Norse who tried their luck in Iceland, or a long, narrow, humid tropical island like Cuba, one with little immediate potential beyond the agricultural and pastoral. Would the resultant culture have been the same? The question answers itself. Or does it? The entity whose origin we seek to explain is a set of basic national characteristics common to all members of a large population inhabiting a land of considerable physical diversity. If there is no one single American physical environment, or even a dominant one, or any sensible possibility of arriving at a meaningful *average* environment, then it is indeed hard to say with what universal physical entities the people in question were interacting to produce a unified national culture, or even that such interaction was truly meaningful. Some escape from this dilemma can be sought in certain pervasive, general qualities of the land that transcend local variability in climate, land forms, soil, and biota. These qualities include sheer vastness, wildness, and plenitude of exploitable natural wealth, and a genuine remoteness from ancestral sources and centers of imperial control. There is also the persuasive, but hardly conclusive, positive correlation between the degree of difference from Britain of the general physical habitat and the regional personality of American culture. Thus, for example, the Southern California cultural pattern seems to deviate from that of Great Britain to a greater degree than does New Hampshire's, roughly to the extent that the Southern California habitat is more unlike the British than is New Hampshire's. Put as simply as possible, it can be asserted, but scarcely proved as yet, that some quite broad physical attributes of the American land—matters of location, bulk, level of resource endowment, but not any specific local habitat—*may* have played a role in the genesis of the *general* structure of American culture. At the same time, regional variants from the national pattern *may* be explainable, in part, in terms of particular local sets of environmental conditions.

d. The appearance of new social and economic patterns, in part in symbiotic response to the local habitat, but more meaningfully perhaps in recognition of deeper forces, promoted cultural change. Although the rules of socioeconomic process and change are embedded in the larger cultural system, when such change does finally occur, it can have some reciprocal impact upon basic cultural structure. That the profound and rapid development of American society and economy did indeed modify many aspects of culture, including some not directly social or economic, and even perhaps some of the basic postulates of the cultural system would seem to be an incontrovertible proposition.

e. Finally, the spatial juxtaposition in America of social and ethnic groups that had been widely separated in the Old World led spontaneously to cultural interchange and the diffusion of old and new ideas and thus must have contributed materially to the forging of the special American identity. In particular, as population grew and internal communications improved, the circulation of messages, goods, and travelers could not help but rise sharply and with a pronounced effect upon cultural patterns. The spatial reshuffling of Old World elements yielded interesting cultural consequences.

5. *A continuing interchange with other parts of the world.* Despite chronic problems in transportation, at least up to the present century, the United States has never been isolated from, or uninfluenced by, the cultural evolution of the remainder of the world.[6] The umbilical cord between immigrant and homeland was seldom totally severed and, in fact, much return migration took place. But beyond such personal ties, the country as a whole has been sensitive to cultural currents elsewhere. This was notably the case during the earlier phases of settlement when the Europeans were restricted to small beachheads, each of which was dependent upon a European metropolis. Later on, with national independence and the shift inland, international cross-pollination was less obvious, though still important. More recently, with a sharp acceleration in the tempo of communications, in the commerce of power, goods, and ideas, and in the circulation of visitors, the impact of the outer world on Americans has grown markedly. At the same time, the effects of American culture on other groups has been magnified even more remarkably.

By dismembering in so clinical—and perhaps pedantic—a fashion, by essentially deductive logic, the confused, untidy narrative of American cultural evolution, we may seem to know more than can be truthfully claimed. In essence, we would like to ask who and what came from where, when, and how, and how they behaved after they got here, in dealing with the land, the people they found on it, and their neighbors

[6] This subject has not yet received the attention it merits, outside the realm of high culture (that is, the fine arts, architecture, letters, and music); but an interesting exploratory attempt is available in Robert O. Mead, *Atlantic Legacy; Essays in American-European Cultural History* (New York: New York University Press, 1969).

near and far. In effect, the preceding paragraphs have outlined an ideal treatment of the historical geography of American culture. Such a monograph is indeed badly needed, and perhaps some day will be feasible; but it is not attempted now, chiefly because our theoretical grasp of the processes of cultural and social change, and of human ecology in general, is still far too primitive to support such an effort. Only fragmentary answers are at hand for any of the queries stated above: it is unlikely that we shall learn very quickly how to analyze fully the ongoing culture changes enveloping us at the moment, much less ever recover enough of the past record to reconstruct unequivocally how Americans came to be the kind of people they are. Furthermore, it is not yet feasible to assign relative weights to the various processes and subprocesses enumerated above. Thus, out of sheer necessity, the basic strategy followed here is descriptive, with only occasional allusions to fundamental process or to larger theory. Nonetheless, although the bulk of this essay is given over to defining sources and end results and to plotting the spatial and temporal contours of our cultural landscapes, with only an occasional boring to the rock-bottom of cultural process, it is in that dark, uncharted realm that we must begin to grope for the larger scientific answers in the not-too-distant future.

Parallels with Other Parts of the World

It cannot be said too often that the means whereby American culture acquired its present form are not peculiar to this part of the world, or that this national pattern is by no means unique. On the contrary, it is useful to classify the United States as a member of a rather large set of countries: the neo-European lands whose culture, population, or both were derived from European sources and implanted successfully at some distance from the homeland, beginning in about 1500 A.D. More specifically, the United States can be viewed as belonging to either of two intersecting subsets of this larger group: a Western Hemisphere area, taking in all of the Americas; or that scattered collection of countries that might be labeled neo-British, namely Canada, Australia, New Zealand, South Africa, Rhodesia, and perhaps Eire, the British West Indies, Bermuda, and the Falklands, in addition to the United States.

The identification of a Pan-American cultural realm is hazardous. Aside from whatever fusion of ideas and behavior might result from a none-too-great spatial propinquity and a certain parallelism in political and settlement history, it is hard to put one's finger on cultural features shared by Anglo-America and Latin America.[7] That there may be an underlying pervasive mood, some exclusive set of shared values, is as yet an open question. Are we the victims of wishful geopolitical thinking? At the very least, however, productive analogies, leaving matters of

---

[7] The strong diversity to be found in Latin America is dealt with in John P. Augelli, "The Controversial Image of Latin America: a Geographer's View," *Journal of Geography*, 62, No. 3 (1963), 105–12.

genuine kinship in abeyance, can be worked out with at least two nations: Brazil and Argentina.

The existence of the neo-British cultural community is too obvious to be argued. Despite great spatial intervals, locations in frequently dissimilar habitats, and envelopment within strikingly alien indigenous and immigrant groups, the family ties and continuities of these sibling countries persist. If ever there were a ready-made laboratory situation for the cultural geographer, here it is. Given two groups of settlers of British background living in two different areas that may, or may not, be physically analogous, how does this or that cultural process develop? Some interesting comparative work has been carried out in the fields of settlement and economic geography,[8] but, rather surprisingly, virtually none aimed directly at cultural issues. Such analyses might be especially worthwhile along the United States-Canadian border, where the only obvious differentials are political and cultural.

The Cultural Legacy of the Old World

We are faced with the problem of describing how much of what was imported into the United States, and to what effect. The matter is immensely complicated by the fact that for the past few centuries Europe has been in a state of swift, even convulsive, change. Adopting a crude metaphor, American immigrants have leaped onto North American soil from a rapidly moving train. Anglo-American culture originated in the very same region that also served as the cradle of modernization. For more than four centuries it has witnessed revolutionary change of a scope, variety, and ultimate planetary impact unmatched in man's million-year existence upon this earth. And if there were profound shifts in the social, political, demographic, scientific, and technological realms, it follows that European cultural patterns were not immune to change.

If we wish, then, to learn what sort of culture the formulators of the American system brought along with them, the current European scene will not be too enlightening. The investigator must resort instead to a rich documentary and archaeological record. Thus, for example, any concern with European antecedents of American folk architecture takes one to old verbal and graphic descriptions and to carefully selected survivals from the past, then to an assessment of basic processes, forms, and meaning—a job, by the way, still waiting to be done properly. A survey of recent construction would be only partially relevant.

A curious byproduct of European cultural dynamism is the fact that

8 For example, John E. Brush and Howard E. Bracey, "Rural Service Centers in Southwestern Wisconsin and Southern England," *Geographical Review*, 45, No. 4 (1955), 559–69; Walker D. Wyman and Clifton B. Kroeber, *The Frontier in Perspective* (Madison: University of Wisconsin Press, 1957); Marvin W. Mikesell, "Comparative Studies in Frontier History," *Annals of the Association of American Geographers*, 50, No. 1 (1960), 62–73, and D. W. Meinig, "A Comparative Historical Geography of Two Railnets: Columbia Basin and South Australia," *Ibid.*, 52, No. 4 (1962), 394–413.

some isolated relics of late medieval or early modern European cultural communities have survived in odd corners of the New World, though suffering extinction in the homeland. Thus a reasonable facsimile of the life of seventeenth-century France or the Rhineland can be glimpsed in veritable living museums in the less accessible tracts of Quebec or amidst the Old-Order Amish of Pennsylvania and other states—and found to be utterly different from present-day France or Germany. Much the same story might be told about survivals of early English and Scotch-Irish traits in the remoter tracts of the Appalachians[9] or about the Volga Germans and Doukhobors. Similarly, the student intent upon learning the pre-Socialist culture of the Polish peasant or the eastern European Jewish *shtetl* would be well advised to try Hamtramck, Chicago, or some isolated North American rural colony for the former and Brooklyn for the latter. Within a wider context, the cultural anthropologist seeking a relatively undisturbed pre-contact West African culture would fare better among the Bush Negroes of the Guyana or Surinam hinterland than among the partially modernized tribes of Ghana or Nigeria.

Even though we cannot decipher too many specific items on that crucial early bill of lading of cultural exports from Europe to America, one outstanding conclusion emerges from fragmentary, indirect evidence: the bearers of this culture were not in any sense a representative sample of the people back home. Instead, we find migrants selected, through external circumstance or by themselves, for specific qualities. The movement of Old World natives to the United States—by all odds the largest migrational shift in human history—was largely spontaneous and uncontrolled. Only in the last few decades, and too late to be of decisive effect, has there been any official screening and rejection of entrants. Although a great mass of public and private documents could be mined to ascertain the kinds of people who decided to hazard their fortunes in America, their motivations, and the processes of selection, to date historians and others have barely begun this monumental chore. However, even the present-day demographer, working with voluminous current data and with living informants who can be tested in great depth, confronts severe problems in measuring motivations or in interpreting differentials in skills, intelligence, and personality traits as between migrants and stay-at-homes or among various categories of migrants. Although selectivity is, and has been, of major import, it need not be simply slanted in one direction, nor is it invariant through time and space.

What, then, was unusual, or nonrandom, about those immigrants who came to be the makers of American culture? First of all, as with almost any modern, long-distance migrational stream, the immigrants were largely young adults, sometimes accompanied by children, and a disproportionate number were male. And, as is also characteristic of most

---

[9] The persistence of Scotch-Irish cultural practices within North America in general is explored in E. Estyn Evans, "The Scotch-Irish in the New World: an Atlantic Heritage," *Journal of the Royal Society of Antiquaries of Ireland*, 95 (1965), 39–49, and "Cultural Relics of the Ulster-Scots in the Old West of North America," *Ulster Folklife*, 12 (1966), 33–38.

contemporary voluntary movements, they came in the main as single families or individuals rather than in larger groups, although clusters of kinfolk or former neighbors would frequently arrive sequentially in a "chain migration." More to the point, certain countries were overrepresented in the passenger lists—Great Britain initially, others later; from within each of the countries in question, regional representation seems to have been quite uneven. This has been well established for seventeenth-century British migration to Massachusetts and Virginia,[10] and is unquestionably just as true for France, Germany, Italy, West Africa, or any other major generator of migrants. Since the degree of cultural diversity among subnational regions was undoubtedly greater during the crucial early colonial period than is the case today, such localization is of more than passing interest. There was probably also some bias in the recruitment of migrants as between urban and rural areas; the city folk may have been overrepresented, but perhaps not consistently.

Too little genuine research has been done on prior education, occupation, or social class of early (or later) American immigrants for any summary statement. Dismiss, if you will, such silly stereotypes as the Virginia Cavalier or the Georgia convict—the former canonized in local myth as the noblest of human material and the latter seen in a quite different light. One plausible generalization is that, contrary to popular belief, poverty was not the chief spur driving the hopeful across the ocean (except, of course, for such panic migrations as followed the Great Potato Famine in Ireland). More probably, a certain critical threshold of incipient affluence, and the appetite for even more, had to be breached before passage was booked. The regional patterns in the historical geography of American immigration seem to support the assumption that simple wretchedness could not do the trick alone. The African immigrant was subject to selective forces beyond his control. The imbalance between sexes among Africans taken as slaves was unusually great, and there was ruthless weeding out for physical stamina. In addition, particularly accessible areas and tribes furnished the bulk of the slave cargo.

We reach much more solid ground in treating the religious composition of the transplanted Europeans. Quite clearly, the immigrants tended to favor the aberrant creeds, whether of the theological left or right. Thus we find gross overrepresentation in colonial America of such nonestablishment groups as the Congregationalist, Presbyterian, Quaker, Mennonite, or Moravian, and perhaps British and German Baptists; and virtually all the French minority in the American colonies were Huguenot. (See Table 1.1.) There is much more to such a strong incidence of churchly dissent than may meet the eye. Adherence to such "fringe" groups no doubt reflected, and reinforced, certain nonmodal personality types and, it is not too bold to assume, deviant political proclivities.

---

10 Charles O. Paullin and John K. Wright, *Atlas of the Historical Geography of the United States* (Washington, D.C. and New York: Carnegie Institution and American Geographical Society, 1932), 46–47, Plates 70–C, 70–D.

Table 1.1.   Religious Congregations: United States, 1775-1776

Congregational	668	Moravian	31
Presbyterian	588	Congregational-Separatist	24
Episcopalian	495	Mennonite	16
Baptist	494	French Protestant	7
Friends	310	Sandemanian	6
German Reformed	159	Jewish	5
Lutheran	150	Rogerene	3
Dutch Reformed	120	Others	31
Methodist	65	Total	3,228
Catholic	56		

Adapted from Charles O. Paullin and John K. Wright, *Atlas of the Historical Geography of the United States* (Washington, D.C. and New York: Carnegie Institution and American Geographic Society, 1932), p. 50.

Would the American Revolution have succeeded, or even have begun, had the voyagers to the New World been a good stratified sample of the originating population?

The most baffling question for the student of migration is that of the intrinsic quality of those involved. How do they rank in terms of native intelligence or various inherent talents? And what about such vague, but important, psychological traits as aggressiveness, achievement drive, gregariousness, docility or its converse, inner-directedness, other-directedness, and a long list of others? All of these are extremely difficult to define and even harder to observe and measure, and are not readily scored on a simple linear scale. Quite possibly the American settlers may have deviated in personality structure to a significant degree, but in complex ways, not always in the same direction, from those they left behind. The existence of a distinctive American national character would indicate either that such an entity reflects the cumulated propensities of just such individuals, who gave American culture its decisive early twist, or that the peculiar conditions of early American life would have led to the same end in any event, or, more plausibly, that some convergence of both these factors took place. That is, a rather special cohort of immigrants interacted with one another and a novel physical and socioeconomic setting to form a strikingly new culture.

The identity of the earlier American immigrants is a question reaching well beyond normal antiquarian curiosity. This is because the colonial record so ideally exemplifies the Doctrine of First Effective Settlement, roughly analogous to the psychological principle of imprinting in very young animals. Whenever an empty territory undergoes settlement, or an earlier population is dislodged by invaders, the specific characteristics of the first group able to effect a viable, self-perpetuating society are of crucial significance for the later social and cultural geography of the area, no matter how tiny the initial band of settlers may have been. As an obvious corollary to this statement, we can ignore the nonviable experiments, for example, the Raleigh group in North Carolina or some ephemeral shore parties in pre-Puritan New England and elsewhere.[11] Thus, in terms of lasting impact, the activities of a few hundred, or even a few score, initial colonizers can mean much more for the cultural geography of a place than the contributions of tens of thousands of new immigrants a few generations later. The history of the northeastern United States clearly illustrates how indelibly the early colonial patterns have marked its cultural landscape.

It is assumed that the culture of the colonial Atlantic Seaboard and, hence, of the later continent-spanning Republic was basically, decisively British. But at least three other distinctively non-British groups were notably active in the colonial arena, and may have contributed to the evolving American culture: the aboriginal, the African Negro, and various continental European ethnic groups. Let us look at each of these in turn.

The Aboriginal Contribution

The aboriginal presence was important for the first wave of pioneers, not only for the early scattered groups along the coast but for most of the later vanguard pressing into inland frontiers as well. Friendly Indians could well tip the balance for physical survival during the very first

---

11 This is as good an occasion as any to note the fascinating riddle of pre-Columbian contacts between the Old and New Worlds. Their relevance to the cultural history and geography of Anglo-America seems much more dubious than may be the case for the Meso-American and South American civilizations. Yet it is increasingly obvious that some voyagers did cross the ocean, perhaps more often by chance than design and not always returning to tell the tale. Indeed it would be extraordinary if such adventures had not occurred. A number of Chinese, Japanese, and other Asian vessels must have drifted to the northwest coast of North America; many undocumented trips between western Europe and eastern North America must have taken place during the fifteenth century or even earlier; the documentary and archaeological evidence for at least intermittent Scandinavian exploration, settlement, and commerce in northeastern North America is irrefutable; the probability of even earlier Celtic voyages is strong; and the possibility of an occasional transatlantic excursion during ancient times cannot be dismissed. But, for our purposes, the net result of all such events is virtually nil. It is barely credible that a few new words were introduced into aboriginal vocabularies and the odd artifact or custom transferred to the Amerindian; but thus far no convincing argument has been advanced for a significant modification of either the pre-Columbian North American or any Old World culture through such casual, shallow contacts. The materials on early transatlantic contacts are summarized and discussed in Gwyn Jones, *The Norse Atlantic Saga* (London: Oxford University Press, 1964); Carl O. Sauer, *Northern Mists* (Berkeley: University of California Press, 1968) and *Sixteenth-Century North America; the Land and People as Seen by Europeans* (Berkeley: University of California Press, 1971); and Douglas R. McManis, "The Traditions of Vinland," *Annals of the Association of American Geographers*, 59, No. 4 (1969), 797–814. The broader range of substantive and theoretical questions concerning early inter-hemispheric cultural contacts is ably explored in a major symposial volume, Carroll J. Reilly, J. Charles Kelley, Campbell W. Pennington, and Robert L. Rands, *Man across the Sea: Problems of Pre-Columbian Contact* (Austin: University of Texas Press, 1971).

years; less amiable tribes might pose major strategic problems if they chose to maneuver against the white settlers on their own or in league with hostile European powers. Nevertheless the sum of the lasting aboriginal contribution to the North American extension of British culture was distinctly meager, and certainly a great deal less than the European impact upon the aborigines.

First of all, a small absolute number of Indians were available to offer their cultural wares, a number that generally dropped sharply during the initial encounters.[12] In many instances, in fact, the tribe simply vanished through abnormally high mortality or flight, leaving the frontiersman with scarcely any aboriginal contacts. This attrition of Indian populations may have been well advanced in many areas before any face-to-face confrontations took place between the two groups. Casual early meetings along the coast probably introduced European infectious diseases that moved rapidly into the interior with highly lethal effect. Equally devastating were European firearms and metal weapons that were carried quickly along well-established trade routes. Thus, for example, when Penn initiated his colony in 1682, it was in a territory already virtually deserted, with the few survivors badly demoralized after nearly a century of indirect onslaught by European maladies and artifacts (and where, incidentally, it was thus prudent to experiment with a humane Indian policy).

A second reason was the unbridgeable cultural chasm between aborigine and European. They differed so profoundly, not only in level of material achievement but in the essential style or shape of cultural pattern that genuine communication was nearly impossible. This basic incompatibility was exacerbated by the spatial accident that British culture was first implanted, and the embryonic American culture gestated, in what chanced to be a backwater region of aboriginal America, the stretch of littoral from the Carolinas to southern Maine. It was late in the colonial era, much too late to redirect the design of a rapidly crystallizing pattern, that American settlers in interior Georgia and the Carolinas first encountered Indian groups whose social organization and level of agricultural technique approached their own. This is quite contrary to the course of events in Spanish America, where the conquerors pushed singlemindedly into the climactic regions of Indian power and culture, expropriated them, then worked outward toward the peripheral zones, and during the process affected a hybridization of both race and culture.

The sharp cultural disconformity between newcomer and aborigine in eastern North America, and later further west, meant a chronic incapacity to formulate a workable social apparatus for accommodating the two disparate groups. To this day, official policy, which once in effect

---

[12] The statistical evidence is meager and grossly defective, especially for eastern North America. The most comprehensive survey of the material—although probably much too conservative in its numerical estimates—is Alfred L. Kroeber, *Cultural and Natural Areas of Native North America* (Berkeley: University of California Press, 1939).

condoned outright extermination, has dithered between the ghetto approach—coralling the survivors into open-air museums called reservations, there to preserve whatever scraps of the ancestral way of life they could—and forced assimilation, that is, transformation into good, copper-skinned WASPs. Both policies have had generally dismal results. No serious attempt has been made, except perhaps in the Hispanic Southwest, to forge either a spatially intermixed, but socially stratified, plural society or a fully, freely fused amalgam of bloodlines and cultural heritages, as has occurred in portions of Latin America, New Zealand, or Hawaii. The point might be one of purely academic interest if the Amerindians would generously agree to vanish, as was thought to be happening around the turn of the century; but now their number is growing at a rate far ahead of that for the general population and one that shows no sign of abating.

What, then, was the Indian contribution?[13] Perhaps the most meaningful gift was a partially humanized land and much valuable information about its contents. The territory entered by explorer and homesteader was not covered by the forest primaeval but had already been grossly modified by aboriginal hunting, burning, forestry, and planting. Therefore much open woodland, or even parkland, was interspersed with fairly large clearings or "Indian old-fields."[14] Indeed in the subhumid regions of the continent, the creation of prairies and grasslands may have been the work of the aborigines. The fact that much practical geographical knowledge was passed on to the new settlers is reflected in the survival of many thousands of place names of Indian origin on the contemporary map. These occur despite the infrequency of Indian words fully integrated into American English, except for those referring to specifically American plants, animals, and land forms. Geographical briefing of the invaders also included data on trails and settlement sites, so that the toil of roadmaker and town founder was somewhat eased. In the case of certain obvious highway routes and town locations, Indian pretesting may simply have hastened the selection.

Considerable knowledge regarding the local biota was handed on to the Europeans. The American pharmacopoeia was immeasurably enriched by Indian lore; and some useful sources of industrial raw materials, such as dyes, fibers, and poisons, were identified by our predecessors. Only a few fully domesticated species seem to have been taken directly by the European settler: maize, various species of beans and squash, the sunflower, and the Jerusalem artichoke. In addition, of course, many other items, some of the utmost importance, were derived

13 For a useful summary, see A. Irving Hallowell, "The Backlash of the Frontier: The Impact of the Indian on American Culture," in *Smithsonian Report for 1958* (Washington, D.C.: Smithsonian Institution, 1959), pp. 447–72.

14 The general prevalence of savannas and other forms of partially open land at the time of the first European incursions into the South, thanks to the activities of the Amerindians, is discussed in Erhard Rostlund, "The Myth of a Natural Prairie Belt in Alabama: an Interpretation of Historical Records," *Annals of the Association of American Geographers*, 47, No. 4 (1957), 392–411.

from tropical America, often by quite roundabout routes, but not from local sources. Among these are: cotton, tobacco, the white potato, the peanut, the tomato, the sweet potato, the turkey, the avocado, and the guinea pig. The local indigenes also shared their knowledge of some semidomesticated and wild food plants, among them the sugar maple, the sassafras, wild rice, the pecan, the persimmon, and a goodly number of fruit, nut, and berry items; and the adoption of various ornamental flowers and shrubs may have been aided by information acquired from the Indians.[15] There are also several minor, but striking, aboriginal items in modern American material culture, notably the canoe, certain forms of basketry, the mocassin, and some types of palisading and crude temporary shelters. No spontaneous transfer at all seems to have occurred in the realms of music, dance, the plastic arts, literature, or religion. Finally, there is the image of the Indian, his mythic role in American literature and folklore.

It might be hypothesized that, had the European colonists found an utterly unpopulated continent, contemporary American life would not have differed in any major respect from its actual pattern. Although some interesting minor nuances can be attributed to interaction with Indians, no basic assumption, attitude, institution, or process seems to have been altered materially thereby. There is the immediate objection: What about the decisive influence of maize in the American farm economy? The fortuitous local presence of this wondrous plant was indeed convenient; but had it been missing, *Zea mais* would surely have been imported from tropical America, probably via Europe, just as soon as a genuine demand for it materialized, precisely as happened with Virginia tobacco shortly after initial settlement, upland cotton from Mexico, soybeans from the Far East, grain sorghum from Africa, or brahmin cattle from India. In any event, Indian corn was totally transformed into a European crop. European, not aboriginal, modes of cultivation were adopted; and it was to play the role first of a bread grain, analogous to wheat or oats, then that of a livestock feed in a manner completely alien to Indian practice, and as a source of completely un-Indian industrial raw materials, for example, mash for bourbon. But if the Indian impact upon Europeans was ultimately negligible, the reverse is certainly not true. When it has endured, the aboriginal culture has been profoundly affected, even fundamentally revised. The now extinct equestrian buffalo-hunting culture of the Great Plains, the Iroquois men who have become so adept at assembling structural steel, and the quite lively Navaho tribe, so strongly focused on sheep (and so adroit in silversmithing) derived from Europe, are only three of the more dramatic examples. Since such battered, but viable, cultures will account for a mounting percentage of the national population, their acculturative practices merit the attention of social scientists and humanists.

[15] U. P. Hedrick, A *History of Horticulture in America to 1860* (New York: Oxford University Press, 1960).

The African Contribution

African Negroes and their descendants, nearly all of them slaves, were present in significant number from 1619 onward and accounted for one-fifth the total population of British North America by the late eighteenth century; and in some parts of the South they greatly outnumbered Caucasians. Under these circumstances, it would be quite extraordinary if some transfer of African culture traits did not occur; but the question of precisely what or how much has not yet been settled at all adequately. A basic reason is that until recently—specifically the publication of Herskovits' deliberately provocative, but scholarly, manifesto[16]—no one had bothered to ask, and that since then much of the work has been partisan in nature. During most of American history, the Negro was invisible to most historians and social scientists, or considered only as the passive occasion for the Civil War. When viewed at all in a social or cultural context, the "obviously inferior" Black was looked down upon as an *Untermensch*, poorly mimicking the master culture. "The myth of the Negro past" is that the slave reached these shores in a condition of utter cultural nakedness or amnesia, quickly jettisoning all of his African heritage or being quite incapable of practising any of it even if he had wished to do so.

The peculiar nature of the slave trade and the process of "breaking in" new slaves did indeed guarantee the obliteration of all but a small residue of the African cultural pattern. Slaves were collected by purchase or capture from much of Black Africa; but the traffic was especially vigorous within West Africa, culturally the most advanced portion of the continent south of the Sahara, except for the Ethiopian Highlands, and clearly the source for most North American slaves. Much mingling of persons from different tribes must have gone on in the ports of embarkation; and if there was an intermediate stopover in the slave markets of the West Indies, further intertribal scrambling took place. This clearly protective tactic of dividing and demoralizing persons who could seldom understand one another's language was carried another step by plantation foremen in the American South. Thus they had at their disposal gangs of slaves who had no choice but to learn a form of English and other facets of the dominant culture when they were thrown together with other detribalized strangers.

It is a marvel that anything at all was retained, but it is plain that a certain, not yet fully specified, residuum of Africanisms has survived. The fact that Africa and Europe, as segments of an Old World in intermittent contact, did share some cultural elements, makes it hard to decide, for example, the source of a particular proverb or folk tale found in America but long current in both Africa and Europe. As in comparable areas in the Antilles and South America, the retention rate for Africanisms is highest in isolated localities where Negroes have been strongly dominant

---

16 Melville Herskovits, *The Myth of the Negro Past* (New York: Harper & Row, Publishers, 1941).

numerically. A prime instance is found in the Sea Islands of South Carolina and Georgia where the Gullah, or Geechee, dialect seems as much West African in syntax and vocabulary as it is English.[17] Something of the same situation may be seen in the Creole speech used by some Louisiana Negroes and introduced from the French West Indies. As in the Caribbean, a vigorous "pidgin" or "Creole" English prevails among American Negroes with little formal schooling. Although formerly denigrated as a degraded or substandard form of English, many linguists now accept the thesis that this form of speech is a legitimate, full-fledged dialect with its own phonetics, vocabulary, and a distinctive grammar, and that African influences were critical in its formation.[18]

It is seldom feasible to relate Africanisms in the American South, unlike those in Cuba or northeastern Brazil, to specific African tribes. But there are some items whose viability in the United States was enhanced by the fact that they transcended tribal lines in Africa. This argument has been made for certain forms of worship and burial practice (including the burial society and methods of grave decoration), and for some types of dance, coiffure, and costume. A very few plants, notably okra, the watermelon, and the peanut (originally, of course, from South America), can be attributed to the Negro slave. There is still much controversy over whether American Negro family structure reflects African tradition or has a later origin. Few African words have been admitted to the standard American English vocabulary, though the score may be better for slang; but there is some possibility that African phonetics may have had their effect upon the general speech pattern of the American South. Musicologists may quarrel endlessly as to just how the blues were born; but it is hard to imagine jazz as having originated without some African inspiration, just as it would equally be far-fetched to think of it springing full-blown out of Africa. It is also the one significant cultural complex of presumable African ancestry, except possibly certain dance forms, to have diffused throughout the entire country.

In other respects, the African cultural contribution, whatever its total extent, was initially limited almost wholly to the Southern culture area. Later, during the post-independence fusion of colonial cultural strains into a national pattern in the trans-Appalachian West, some traces of African culture would automatically have been included in the Southern parcel of traits, and thus might appear in quite dilute form in the national culture. The recent self-conscious revival of African practices by the more determined Black Militants—notably in costume, coiffure, and personal names, with some discussion of adopting African languages—is extremely interesting in its own right as another example of the deliberate

[17] Lorenzo Dow Turner, *Africanisms in the Gullah Dialect* (Chicago: University of Chicago Press, 1949).

[18] William A. Stewart, "Continuity and Change in American Negro Dialects," *The Florida FL Reporter* (Spring 1968). Bibliographic citations for much of the useful literature on this intensely controversial topic are to be found in this provocative essay.

manipulation of symbols in synthesizing ethnic self-awareness; but these ultra-visible gimmicks are items from the library or the travelogue, not treasures dug up from the cellar of folk memory. ·Those few Negro Americans who have lived or visited in Africa for any extended period are quickly, acutely made aware of the profound cultural gulf between themselves and their racial brethren. Similarly, there is a sharp separation in both physical and cultural space between the repatriated Americanized Negroes of the Liberian coast and the indigenous communities of the interior.

The Continental European Contribution

As they solidified their hold upon the Atlantic Seaboard from Maine to Georgia, the British (that is, the English, plus the Anglicized Welsh, Cornish, and Scotch-Irish) willy-nilly acquired title to an appreciable number of non-British settlers, namely some antecedent Dutch, Flemings, Swedes, and Finns in the Hudson and Delaware valleys and vicinity. In addition, large numbers of Germans and ·Swiss began entering Pennsylvania and Maryland from about 1700 onward. Further ethnic spice was added elsewhere by smaller contingents of Highland Scots, Irish Catholics, Sephardic Jews, and French Huguenots. With the great territorial expansion of the American Republic, a major community of French-speaking folk was annexed in southern Louisiana and smaller clusters elsewhere in the Mississippi Basin and the Great Lakes country. Similarly, a goodly number of persons of Spanish and Mexican stock suddenly found themselves within the borders of the United States in 1848. Finally, the Russian presence had made itself felt, however fleetingly, in northern California as well as in Alaska. How great a contribution did these varied non-British European groups make toward the formation of the American culture, regionally or nationally?

In the case of the colonial hearth areas, that is, the nodal zones along the Atlantic Seaboard that proved to be the incubators for the national cultural pattern, there is some uncertainty as to whether the Doctrine of First Effective Settlement applies to the few Swedes, Finns, and Dutch along the Delaware Valley. These settlers and their way of life were promptly swamped by subsequent British and Teutonic colonization. The Dutch were more successful in clinging to some shreds of identity for more than two hundred years in Manhattan, Long Island, New Jersey, and the Hudson Valley; but eventually they too were smothered by the greater weight of British, or general American, population and culture. Some lingering traces of their cultural presence may be noted locally, but any contribution to national culture is problematic. Similarly, such small, exotic enclaves as the Salzburgers in Georgia, colonies of Swiss and Huguenots in South Carolina, or Highland Scots in North Carolina were quite local and ephemeral as regards the genesis of national patterns.

The case of the so-called Pennsylvania Germans is more complex. This appellation fits a rather heterogeneous set of German-speaking

Protestants from Switzerland and the Palatinate and other sections of Western Germany. They began arriving in force in Pennsylvania and Maryland quite soon after English-speaking settlers had firmly occupied these tracts, and were thus just barely among the first effective settlers. Soon thereafter, the Pennsylvania Germans formed a significant part of the frontier contingent in the Appalachian Piedmont and valleys from Virginia through Tennessee, and were among the first groups to penetrate the eastern Middle West. This ethnic group may account for a major share in culture formation within an influential area centering on the Delaware and Susquehanna valleys, and, by means of migration and diffusion, may have contributed consequentially to the national mix. The question is muddied by the fact that, within the Old World, the British, Germans, and neighboring folk were co-members of a cultural subrealm with rather efficient internal communications.[19] Thus it is usually difficult to assign a particular trait or innovation to a specific source. Take the introduction of log-building techniques in house and barn construction. This may have been the work of the Pennsylvania Germans, but an equally strong case could be pleaded for the seventeenth-century Swedes and Finns.[20] In any case, the Pennsylvania Germans—and, with major reservations, the African Negroes—would appear to be the only credible nominees as non-British participants in the critical early phases of creating a national American culture.

The fate of the annexed non-British ethnic groups is of some interest, at least locally. The Florida Spaniards and California Russians can be written off immediately, for they quickly sank out of sight. The tiny nodes of French settlement in the Mississippi-Great Lakes territories (for example, Detroit, Michilimackinac, Prairie du Chien, Vincennes, Kaskaskia, Ste. Genevieve, and the eastern Ozarks) succumbed only a little less rapidly, leaving as their lasting legacy only a few place names and an esoteric land survey geometry. Despite much romantic (or commercial) self-deception, the same story must be told for the Colonial Spanish and Mexican cultures in California. Their contribution, especially in the agricultural sector, might have persisted more strongly had the transition to a new regime been smoother; but the antecedent cultures were simply pulverized under the massive Anglo-American avalanche, leaving little more than a few ruins, place names, and archaic grant boundaries, and some items of agriculture and animal husbandry.[21]

Genuine survival and durability are found in French Louisiana and particularly in the Hispanic Southwest. The Cajuns, speaking a French patois, have managed, in part because of relative isolation, to cling to

[19] The point is argued strongly in James T. Lemon, *The Best Poor Man's Country; a Geographical Study of Early Southeastern Pennsylvania* (Baltimore: The Johns Hopkins Press, 1971).

[20] The problem is discussed in Harold R. Shurtleff, *The Log Cabin Myth* (Cambridge, Mass.: Harvard University Press, 1939) and Terry G. Jordan and Matti Kaups, *The American Backwoods Frontier* (Baltimore: The Johns Hopkins Press, 1989).

[21] H. F. Raup, "Transformation of Southern California to a Cultivated Land," *Annals of the Association of American Geographers,* 49, No. 3, Part 2 (1959), 58–78.

their linguistic, religious, and, to a degree, political identity, so that despite a powerful dose of Americanization and a certain culturally schizoid tendency in consequence, they retain their ethnicity, and may do so indefinitely. In New Mexico and southern and southwest Texas, and some segments of Arizona and Colorado, the political and economic supremacy of the Anglos and a strong recent in-movement from the East have not yet fully overcome the cultural stubbornness and demographic vigor of an Hispanic or "Chicano" (and in New Mexico a partially Hispanicized aboriginal) population.[22] Their staying power is abetted by the continuing immigration and circulation of Mexicans and a constant flow of cultural currents from the neighboring motherland. The Hispanic Southwest remains a battleground of cultures, the major example of cultural pluralism in the United States, and, in some respects, an accidental political (and ecological?) protrusion of the Anglo-American realm beyond its logical limits into the periphery of Latin America. In any case, the strong non-British heritage of the Hispanic Southwest, like the Gallic tradition of southern Louisiana, is a local affair. With the possible exception of the range livestock system and the attendant "cowboy complex,"[23] little has been exported past regional boundaries; the mainstream of national culture is unaffected.

Significance of Sequent Waves of Immigration

The chronicling of American immigration and a discussion of the major processes at work are most pertinent to any analysis of American cultural geography.[24] The sequential ordering of unequal parcels of ethnic groups was a material factor in the development of the national culture. Their initially irregular spatial assignment promoted regional distinctiveness among culture areas; and the persistence of ethnic areal segregation is, of course, a topic of consuming interest in and of itself.

A survey of the more than 360 years of foreign immigration into

22 Richard L. Nostrand, "The Hispanic-American Borderland: Delimitation of an American Culture Region," *Annals of the Association of American Geographers,* 60, No. 4 (1970), 638–61; and Donald W. Meinig, *Southwest: Three Peoples in Geographical Change* (New York: Oxford University Press, 1971).

23 Walter Prescott Webb, *The Great Plains* (New York: Grosset & Dunlap, 1957), 205–15. For a dissenting viewpoint, see Terry G. Jordan, "The Origin of Anglo-American Cattle Ranching in Texas: a Documentation of Diffusion from the Lower South," *Economic Geography,* 45 (1969), 63–87.

24 The voluminous literature on American immigration is listed in Stanley J. Tracy, ed., *A Report on World Population Migration as Related to the United States* (Washington, D.C.: George Washington University Press, 1956). Among the more outstanding treatments of the subject are: Oscar Handlin, *The Uprooted: the Epic Story of the Great Migrations that Made the American People* (New York: Grosset & Dunlap, 1951); Marcus Lee Hansen, *The Atlantic Migration, 1607–1790: A History of the Continuing Settlement of the United States* (New York: Harper & Row, Publishers, 1961); Conrad Taeuber and Irene B. Taeuber, "The Immigrants," in *The Changing Population of the United States* (New York: John Wiley & Sons, Inc., 1958), pp. 48–70; and Maldwyn Allen Jones, *American Immigration* (Chicago: University of Chicago Press, 1960).

the territory of the United States suggests a division into five distinct periods, as follows:

    I. 1607–1700: The initial, strongly English and Welsh wave, along with a major complement of Africans;
   II. 1700–1775: A still predominantly English, Welsh, and African movement, but with strong Teutonic and Scotch-Irish components as well;
  III. 1820–1870: The Northwest European wave—heavily British, Irish, Teutonic, and Dutch, but including the vanguard of other European groups and some Asians, Canadians, and Latin Americans;
  IV. 1870–1920: The Great Deluge—large numbers of eastern and southern Europeans and Scandinavians joining those from Northwest Europe; Asians, Canadians, and Latin Americans in significant quantity;
   V. 1920 to present: A highly miscellaneous influx—less than in the preceding period but still of major absolute magnitude—from western and southern Europe, Latin America, Canada, and some Asian lands, with Latin America registering especially steady and major gains.

Firm quantitative values cannot be assigned to these periods. Official statistics are lacking prior to 1821; and, since then, many illicit entries have eluded detection. Furthermore, the official figures indicate gross rather than net immigration, for a considerable percentage of the foreign-born do eventually leave the country, and there is an outward trickle of native Americans as well. If a rough guess must be hazarded for immigration before 1820, then a net total of far less than one million seems plausible.

The cultural impact of these five periods is inversely related to their recency, in keeping with the Doctrine of First Effective Settlement. Thus the few tens of thousands who arrived before 1700 were much more consequential for our cultural heredity than the 26,240,000 tabulated for the 1870–1920 period. To adopt a somewhat fanciful biological analogy, Period I could be likened to cultural infancy, Period II to childhood, and the last three periods to adolescent and mature development. Behaviorally, the earlier formative years are the important ones. Carrying the metaphor even further, it must be admitted that the act of conception in western Europe was probably the most critical event of all; but its exegesis lies beyond the limits of this essay.

These five periods can be grouped into two larger eras that are quite distinct in terms of the acculturation process: a colonial period up through 1775, and a national period following 1820. During the first, the migrant gave no thought to cultural adjustment, whatever other problems may have beset him. He left Great Britain with the realistic expectation of remaining very much the same sort of Englishman he had always been, for acculturation was subtle and painless. Even the former residents of the Rhineland planned to preserve their identity; and after settling in fairly cohesive territorial blocks, most did succeed, as have many of their desendants after some two and a half centuries. Those who came after the resumption of large-scale immigration in the 1820's were faced with a new situation: a dynamic, evolving national culture already sharply divergent from its European forebears. Active membership in it meant paying heavy dues: education in a new language or dialect and indeed learning a whole new way of life, even for the British newcomer. In brief, immigrants experienced "culture shock" followed by total acculturation, if not in the first generation, then certainly in the second or third. It could not have been very long before the immigrant began to anticipate the ordeal. The fact that so many millions of the foreign-born and their children agonized their way through a prolonged personality crisis deeply compounded the chronic American dilemma. For after they had paid the stiff psychological entrance fees, the "new Americans" had joined a club whose older members were and are forever searching out their elusive identity, consciously trying to decide what the new land and the new man are to be.

Exemption has been granted only to those immigrants who were not forced to snap the ties with the homeland abruptly: the French-Canadians, Mexicans, and Puerto Ricans. Those French-Canadians filtering into New York and New England from Quebec and New Brunswick, but still commuting socially and psychologically, if not physically, between domicile and homeland, have not yet all been Americanized past the point of no return. Exactly the same situation prevails for those Mexicans crossing northward into Texas and the Southwest in general. The tether binding Puerto Ricans and Virgin Islanders to their home islands is a short one, in hours and dollars, if not in miles; and their comings and goings are frequent enough to subvert acculturative tendencies.

That the first effective wave of immigrants (1607–1700) was so largely British (or, in the case of Canada, French) is one of the pivotal facts of North American human geography. The exploitation and settlement of eastern North America by Northwest Europeans—primarily the British, of course, but, in descending order, the French, Dutch, and Swedes as well, were not quite as inevitable as may now appear. For more than a century, the Spanish failed to exercise seriously their military and legalistic option on North American real estate, confining themselves to a few flimsy coastal missions in Georgia, the Carolinas, and Chesapeake Bay, and their pathetic outposts in Florida. Several excuses can be cited: lack of sufficient manpower and naval strength; a general preoccupation with the administration of New Spain; the poor missionary prospects offered by the North American aborigines. The basic reasons, however, were probably that the Spanish found the region physically and economically uninviting and, in any case, had not yet attained that stage in their internal development when large numbers of emigrants with capital could be ejected from the homeland. The British found eastern North America a much more attractive place than the Spanish did and by 1600 were economically, demographically, and otherwise primed for foreign adventure and the spawning of a series of new Englands. After a century and a half of combat with like-minded continental northwestern Europeans, the British swept eastern North America clear of competition by virtue of superior military, institutional, and economic strength.

The Seventeenth-Century British immigrants were drawn from most sections of the island of Great Britain (and in much smaller number from the British West Indies). A great deal of interregional scrambling took place in the assembling, trans-Atlantic passage, and subsequent interactions of the travelers, so that original distinctions were blurred or totally erased. Only the most deviant group, the Welsh, managed to hang on to much of their parochial identity, and then only temporarily, as in parts of southeastern Pennsylvania. In effect, a sort of basic Pan-British kit of culture traits survived and was implanted, in precisely parallel fashion to what has been noted for New Spain, where a basic Pan-Hispanic pattern was transmitted, in part through formalized edicts and codes, to all corners of a vast empire. The result was unmistakably British, but regional contributions could no longer be sorted out.

The eighteenth-century wave of immigrants witnessed greater ethnic diversification with the arrival of shiploads of the quasi-British Scotch-Irish, the quite un-British Teutons, and a trickle of others. Even earlier, in Philadelphia and Nieuw Amsterdam-New York and their respective hinterlands, the ethnic and religious mix was more varied than elsewhere. Small Dutch, Flemish, Jewish, French, Spanish, and Italian groups were to be encountered, in addition to the substantial communities of Germans and both slave and free Negroes. Up until the suspension of major movements in 1775, the emigrant ships disgorged their cargoes mainly within a limited number of harbors from Portsmouth, New Hampshire to Hampton Roads (movement inland from the Carolina and Georgia coast was feeble). The newcomers fanned out northward, inland, and southwestward in a great, irregularly pulsating arc.

After more than four decades of military disturbances in both America and Europe, significant immigration resumed around 1820 and, with momentary quiet periods caused by war or economic depression, increased until an all-time peak was reached in the opening years of the twentieth century. The veritable flood that had poured across the ocean for nearly a century was finally reduced to a modest trickle by World War I. It is convenient to divide this great mass movement into two fifty-year periods, before and after 1870. During the first, natives of Great Britain were joined by an extraordinary outpouring of famine-stricken Irish, swarms of Germans, and lesser, but significant, groups of Swiss, Netherlanders, and Belgians (but surprisingly few French). Since the importation of Africans had been outlawed in 1808 and the contraband slave trade was minor, this phase can be fairly characterized as Northwest European.

From about 1870 onward, the flow from Northwest Europe continued but was augmented dramatically by immigration from Scandinavia and from eastern and southern Europe. Poles, Hungarians, Lithuanians, Jews, Russians, Ukrainians, Czechs, Slovaks, Croats, Serbs, Italians, Macedonians, Portuguese, and Greeks were all abundantly represented. In addition, the influxes from Canada, Latin America, China, and Japan, already in progress before 1870, gained in volume.

Viewed at the macroscopic scale, Europe's development moved steadily outward in wavelike fashion through time and space from a hearth area around the North Sea toward the north, east, and south; and it is clear that the capacity and impetus for intercontinental migration was a major symptom of the achievement of a certain significant level of advancement. (See Table 1.2.) In fact, one can rather accurately date the inception of significant modernization for individual nations by close study of emigration data.[25] A mirror image of the migratory impulse in the European subcontinent was the spatial zonation of ethnic groups in the United States and Canada, especially within the rural population. (As time went on, immigrants gravitated toward the cities at a greater rate even than native Americans.) The late arrivals were more likely to seek opportunity in the open spaces of the central and western United States than in the eastern third of the country, where the more inviting rural niches had been preempted. And there is indeed a special concentration of Germans in the upper Middle West, of Scandinavians in the north central states, Italians in California, or of various Slavic groups from Kansas northward and northwestward.

As the settlement frontier rolled westward, there seems to have been a gradual change in the size of ethnic blocs. Within the original colonies, large blocks of land were dominated by single groups—whole counties or even provinces. Thus we find that, formerly at least, almost all of New England had much the same ethnic composition, as did much of the South. Even in heterogenous Pennsylvania, specific groups would tend to dominate specific counties. At a later stage, during the settlement of the Middle West, rural immigrant blocs seldom prevailed beyond the township scale, giving rise to intricately meshed mosaics in such states as Wisconsin or Kansas;[26] and each major city might have several distinct ethnic neighborhoods. In the final stage, in the Far West, large uniform areas of any single European group have been rare; within the city, group dominance seldom goes beyond the single residential block, and one block may hold several nationalities. A similar pattern is found in the western countryside. Evidently, there has been a growth through time, and thus with distance westward, in the spatial porosity of American society, possibly because of a lessening need for territorial clannishness, but more probably because of the nature of land law and private land speculation in recent decades.

In the final, current phase of immigration, both the volume and national composition of the movement are controlled by federal statute. But even without such official restriction, the flow would have ebbed. The American "pull," especially for the relatively unskilled laborer, had become weak by the 1920's, and so had the economic (if not sociopolitical) "push" to abandon Europe. In any case, immigration from Latin America, largely uncontrolled, has grown rapidly and is now clearly the largest regional category. Although present-day immigration has had

---

[25] The phenomenon is mapped in *Atlas of the Historical Geography of the United States,* Plates 70–E through 70–O, and discussed in Donald J. Bogue, *Principles of Demography* (New York: John Wiley & Sons, Inc., 1969), 804–10.

[26] J. Neale Carman, *Foreign-Language Units of Kansas. Historical Atlas and Statistics* (Lawrence: University of Kansas Press, 1962).

Table 1.2. Immigrants to the United States, 1820-1969, by Region or Nation of Origin and Percent of Decadal Total

	Northwest Europe				Germany	Italy	Spain, Portugal and Greece	Eastern Europe[b]	U.S.S.R. and Baltic States	Asia	Canada	Latin America[c]	All Other Countries	Total No. (in thousands)
	Great Britain	Ireland	Scandinavia	Other NW Europe[a]										
1820–1829	20.5%	40.2%	0.2%	9.3%	4.5%	0.3%	2.1%	—	0.1%	—	1.8%	5.8%	15.2%	129
1830–1839	13.8	31.7	0.4	8.4	23.2	0.4	0.5	0.1	—	—	2.2	3.7	15.6	538
1840–1849	15.3	46.0	0.9	6.4	27.0	0.1	0.1	—	—	—	2.4	1.1	0.5	1,427
1850–1859	15.8	36.6	0.9	4.4	34.7	0.3	0.4	0.1	—	1.3	2.3	0.7	2.7	2,815
1860–1869	25.6	20.5	4.6	3.4	34.7	0.5	0.4	0.2	0.1	2.6	5.7	0.6	0.9	2,081
1870–1879	21.1	15.4	7.6	4.3	27.4	1.7	0.7	2.6	1.3	4.9	11.8	0.8	0.4	2,742
1880–1889	15.4	12.8	12.8	3.8	27.6	5.1	0.4	6.9	3.5	1.3	9.4	0.6	0.3	5,249
1890–1899	8.9	11.0	10.6	3.3	15.7	16.3	1.3	17.7	12.2	1.6	0.1	0.9	0.4	3,694
1900–1909	5.7	4.3	5.9	2.2	4.0	23.5	2.9	26.3	18.3	2.9	1.5	1.9	0.6	8,202
1910–1919	5.8	2.6	3.8	2.5	2.8	19.4	5.7	19.9	17.4	3.1	11.2	5.7	0.3	6,347
1920–1929	7.9	4.8	4.7	3.2	9.0	12.3	3.8	12.1	2.1	2.6	22.1	14.9	0.4	4,296
1930–1939	7.7	5.1	2.4	4.6	17.1	12.2	3.3	10.0	1.3	2.6	23.3	9.5	0.8	699
1940–1949	14.6	2.6	2.6	8.7	13.9	5.9	2.4	3.9	0.5	3.5	18.8	19.5	2.9	857
1950–1959	12.4	2.4	2.1	6.3	24.6	8.0	2.8	5.6	0.1	5.7	15.3	12.6	1.8	2,300
1960–1969	7.1	1.3	1.4	3.7	6.5	6.2	6.0	3.6	0.1	11.0	13.5	38.9	1.1	3,212
Total No. (in thousands)	4889	4714	2,473	1,635	6,896	5,149	1,155	5,384	3,387	1,431	3,942	3,087	463	44,588
	10.9%	10.6%	5.6%	3.7%	15.2%	11.5%	2.6%	12.1%	7.6%	3.2%	8.9%	6.9%	1.0%	100.0%

a Belgium, Netherlands, Luxembourg, France, and Switzerland.
b Albania, Austria, Bulgaria, Czechoslovakia, Hungary, Poland, Rumania, Turkey in Europe, and Yugoslavia.
c Not including persons born in Puerto Rico, Virgin Islands, or Canal Zone.
Adapted from Donald J. Bogue, *Principles of Demography* (New York: John Wiley & Sons, Inc., 1969), p. 807 and *Statistical Abstract of the United States*, 1969 and 1970.

far less impact, qualitatively or quantitatively, than that of the past, the admission each year of 500,000 to 700,000 persons is greatly in excess of the number reported by any other nation.

In no case, with the possible exception of the French-Canadian invasion of Aroostook County, Maine and the northern segment of Vermont or the sudden Cuban surge into Miami, has foreign immigration displaced or fundamentally reordered the American culture. Nevertheless, it may have brought about major modification. Thus large parts of Minnesota and the Dakotas are unmistakably Scandinavian and American; the Germanic character of portions of south central Texas[27] and the Dutch imprint upon southwestern Michigan are plain to see,[28] as is the recent Jewish-American characer of the Catskills; and much of Pennsylvania's and West Virginia's coal belt has been markedly slavicized. The personalities of several major metropolises have been deeply affected by large ethnic settlements. The Germans of Milwaukee and Cincinnati, the Poles of Detroit and Buffalo, the Irish of Boston, the Puerto Ricans in New York City, and the Cubans and Jews of Miami come immediately to mind.

Territorial Patterns Among Ethnic Groups

The whole grand theme of the ethnic diversity of the nation, considered in its full spatial range, from the major region down to the fine, intraurban details, merits much more attention than it can be given here.[29] We limit ourselves here to suggesting a few major processes that

27 Terry G. Jordan, *German Seed in Texas Soil; Immigrant Farmers in Nineteenth-Century Texas* (Austin: University of Texas Press, 1966).

28 Elaine M. Bjorklund, "Ideology and Culture Exemplified in Southwestern Michigan," *Annals of the Association of American Geographers,* 54, No. 2 (1964), 277–41, and John A. Jakle and James O. Wheeler, "The Changing Residential Structure of the Dutch Population in Kalamazoo, Michigan," *Ibid.,* 59, No. 3 (1969), 441–60.

29 It is rather surprising how little attention has been devoted, at the national scale, to the ethnic-racial composition of the American composition, as viewed in spatial perspective. The most recent cartographic treatments are: S. I. Bruk and V. S. Apenchenko, eds., *Atlas Narodov Mira* (Moscow: Main Administration of Geodesy and Cartography, 1964), 94–95, (a map depicting no less than 54 different groups); and *Atlas of the Historical Geography of the United States,* Plates 67B–70B, 71–76A. The textual notes for the latter (pp. 46–48) provide references to earlier census atlas coverage of immigrant stock. There is a brief but useful overview of the broader geographic patterns in Richard Hartshorne, "Racial Maps of the United States," *Geographical Review,* 28 (1938), 276–88, and a more extensive, but only intermittently geographic treatment in Donald Keith Fellows, *A Mosaic of America's Ethnic Minorities* (New York: John Wiley & Sons, Inc., 1972). A number of studies dealing with individual groups, mostly at the state or local level, are listed in Wilbur Zelinsky, *A Bibliographic Guide to Population Geography,* Department of Geography, Research Paper No. 80 (Chicago: University of Chicago, 1962). The recent literature on Afro-Americans is listed in Donald R. Deskins, Jr., "Geographical Literature on the American Negro, 1949–1968: a Bibliography," *Professional Geographer,* 21, No. 3 (1969), 145–49. Among the more interesting post-1961 items analyzing ethnic geography at the micro scale are Daniel F. Doeppers, "The Globeville Neighborhood in Denver," *Geographical Review,* 57, No. 4 (1967), 506–22; and Gerald D. Suttles,

may have helped create the territorial zonation of ethnic groups. In the first instance, there is the distance-decay effect, or the importance of relative proximity. This has already been shown to operate for the French-Canadians of the Northeast and the Spanish-Americans of the Southwest. It is also the principal factor accounting for the Pacific Coast concentration of the Chinese and (especially before World War II) the Japanese (Figure 1.1) as well as the lesser number of Filipinos, Koreans, and East Indians, and for Cubans in Florida and along the Gulf Coast. Less obviously, this consideration governed the locational choices of the colonial British, Germans, and Scotch-Irish, so that often nothing more mysterious than the question of the nearest available real estate was involved. Spatial friction also explains, in part, the great piling up of nineteenth- and twentieth-century immigrant groups in the major ports of entry.

With much trepidation, it can be hinted that the factor of environmental affinity—the selection of familiar and preferred kinds of habitat—might bear looking into. Certainly other causes played their part in the heavy Scandinavian concentrations in the North Central states; but the availability of terrain, plant cover, soil, and, less convincingly, climate reminiscent of the homeland or responsive to known skills is to be considered. This is particularly true for certain marginal tracts claimed by Icelanders and Finns, which were ignored by members of a less hardy breed. Similarly, the success of Dutch farmers in polderizing marshlands in Michigan and Ontario supports the hypothesis, as may the flocking together of Italians and Armenians in sections of California's Central Valley that are crudely analogous to their motherlands.[30] Environment may be involved only obliquely, as when specific groups are introduced because of occupational specialization. Thus Basque shepherds are concentrated in various parts of the Intermontane West, Japanese horticulturists in Southern California, Portuguese fishermen around Cape Cod, Greek sponge-divers in the Tampa area, and Cornish miners in southwestern Wisconsin. And there is the overwhelming geographic fact of a large Negro population corresponding to the plantation system. Other ethnically linked occupations, however, are only weakly place-specific: Jewish peddlers and merchants, Greek restaurateurs, and Chinese laundrymen.

Once a viable ethnic nucleus takes hold at a given location, chain migration may be triggered. If communication lines are kept open between the new settlements and relatives and neighbors back home, positive information may induce the latter to pack up and follow. In this way, a great many urban and rural ethnic neighborhoods have expanded. This mechanism has also reinforced the ranks of at least one major religious

The Social Order of the Slum; Ethnicity and Territory in the Inner City (Chicago: University of Chicago Press, 1968).

   [30] The location of the Italian-American population is presented and discussed in Joseph Velikonja, "Distribuzione Geografica degli Italiani negli Stati Uniti," *Atti del XVIII Congresso Geografico Italiano, Trieste, 4–9 Aprile 1961* (Trieste, 1961), 1–20.

1-a
AFRO-AMERICAN

○ Primary metropolitan concentration
• Secondary metropolitan concentration

1-b
JAPANESE & CHINESE; SCANDINAVIAN

Scandinavian rural concentrations
Japanese & Chinese rural concentrations

	Primary metropolitan concentration	Secondary metropolitan concentration
JAPANESE & CHINESE	□	○
SCANDINAVIAN	×	•

1-c
AMERINDIAN

Secondary metropolitan concentration

1-d
GERMAN

○ Primary metropolitan concentration
• Secondary metropolitan concentration

■ Primary rural concentration   ▓ Secondary rural concentration

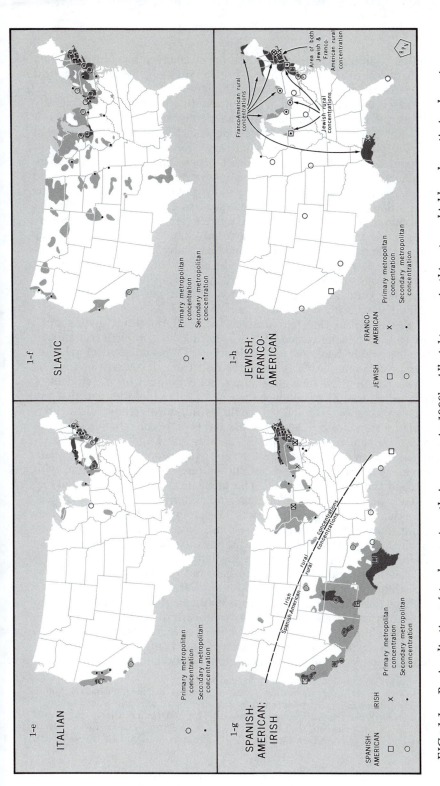

FIG. 1.1. *Regionalization of twelve major ethnic groups in 1960's. All eight maps in this set are highly schematic in character. Since a specific sliding scale, one combining absolute and relative values, was devised for each, and since the definition of ethnicity and the quality of data vary sharply from one group to another, it is not advisable to make quantitative comparisons between any pair of maps, or portions there of.*

1-e
ITALIAN

○ Primary metropolitan
   concentration
• Secondary metropolitan
   concentration

1-f
SLAVIC

○ Primary metropolitan
   concentration
• Secondary metropolitan
   concentration

1-g
SPANISH-
AMERICAN;
IRISH

SPANISH-   IRISH
AMERICAN
□   ×   Primary metropolitan
                       concentration
○   •   Secondary metropolitan
                       concentration

Irish rural concentrations
Spanish-American rural concentrations

1-h
JEWISH;
FRANCO-
AMERICAN

Franco-American rural concentrations
Jewish rural concentrations
Area of both Jewish & Franco-American rural concentration

JEWISH   FRANCO-
              AMERICAN
□   ×   Primary metropolitan
                       concentration
○   •   Secondary metropolitan
                       concentration

group transcending ethnic lines: the labors of nineteenth-century Mormon missionaries abroad inspired many British, German, Scandinavian, and other converts to make their way to Utah.

There is one large section of the country that has failed conspicuously to attract the foreign-born Caucasian in significant numbers: the American South. Even before the Revolution, the South had been peopled largely by pioneers trekking overland from Pennsylvania, Maryland, and the Upper South rather than by direct penetration of its coastal plain. Subsequently, all but a minute fraction of the vast immigrant horde avoided Southern destinations. The only notable regional exceptions are a portion of Texas and (if it is indeed southern) southern Florida. The apparent explanations are: an inferior array of economic opportunities, even after Emancipation; a social climate that few immigrants found enticing; and a xenophobia that may have been both cause and effect of the low incidence of aliens. In any event, the absence of any immigrant leavening of the solid WASP-cum-Afro-American regional culture of the South affords some negative insight into the nature of the immigrant impact upon the North and West.

The validity of that perennial cliché, "the American Melting Pot," merits at least passing notice here, for, in fact, the problem has already spawned a whole bookshelf of publications.[31] The question of whether felt ethnic differences will gradually vanish from the American scene is obviously one of great theoretical and practical import for the student of culture or society. The confident forecast of a few decades ago that the powerful forces of assimilation would grind down ethnic distinctions and mold a standardized American has not been borne out. Some foreign groups do assimilate rather rapidly, notably the English, British Canadians, and many Scandinavians; in fact, the propensity to assimilate seems to be a function of the specific culture—and of generation, with the second the most susceptible. The phenomenal territorial mobility of Americans and a high rate of intergroup marriages also help to tear down ethnic and religious barriers. Yet, despite the obvious forces pointing toward ethnic entropy, the sense of difference and affinity for particular places shows no signs of disappearing in either rural neighborhoods or urban centers. Indeed, in the latter case, some ethnic strongholds have been staked out in suburban and exurban tracts as well as kept within the inner cities. The Melting Pot, it turns out, contains a lumpy stew; and though ethnic groups may have thickened and enriched the All-American broth—except in the southeastern quadrant of the pot—the lumps will not cook away, at least not for a good many more years. We can now expect diversity to persevere for many years. It might be added that we lack any serious geographic analysis at the local scale of

---

31 Two of the more interesting recent approaches to the problem are Milton M. Gordon, *Assimilation in American Life: the Role of Race, Religion and National Origins* (New York: Oxford University Press, 1964); and Nathan Glazer and Daniel Patrick Moynihan, *Beyond the Melting Pot: The Negroes, Puerto Ricans, Jews, Italians, and Irish of New York City,* 2d ed. (Cambridge, Mass.: M.I.T. Press, 1970).

the processes of ethnic assimilation or the lack of it in the United States and that such efforts might be exceedingly useful.

How great was the effect of the post-1820 immigration upon the character of the American cultural system or its geography? The general answer must be: quite considerable in outward detail, rather little in inner substance. By the year 1820, the essential business of modeling a new national culture and consciousness with its special set of assumptions, drives, and rules had been completed. On a superficial level, it is easy to comment on how our lexicon has been enlivened by additions from the German, Irish, Mexican, or Yiddish vocabularies or how our menus have been similarly enriched by immigrant cuisine. On a national scale, the exotic addenda of the past century and a half have diluted certain characteristics and strengthened others; but none of the central defining attributes, as set forth in the next chapter, has been revised. Perhaps the major effect has been to reinforce something already in progress, the move from an English to a native American ethos, despite our continuing "special relationship" with Britain. An ecumenical ingathering of nations has given substance to the dream of a brotherhood of common ideals and virtues open to all (but most cordially to Caucasians) rather than a shared ancestry—of a voluntary utopia in place of the doctrine of blood and soil.

The Turner Thesis

No discussion of how American culture originated could be complete without treating the most popular, and certainly the most vigorously debated, of the general hypotheses set forth to date, that of Frederick Jackson Turner, first enunciated in 1893, and further elaborated in later writings by Turner and his disciples.[32] Ignoring the many interesting tangential questions that spring from it, the essential argument is quite simple: The central fact of American existence has been the experience of the advancing settlement frontier and its effects upon the rest of the country and later generations. Turner maintained that the American national character was forged on the frontier and that the meaning of American history and institutions can be read only in the light of the peculiar social, economic, and psychological conditions of that transient zone. The lively controversy that has raged between the strongly polarized proponents and opponents shows little sign of abating even now, with only a few scholars working toward a compromise; but it is only in such an intermediate position, I believe, that the merits of both sides of the argument can be realized.

[32] The essential utterances can be found in Frederick Jackson Turner, *Frontier and Section: Selected Essays* (Englewood Cliffs, N.J.: Prentice-Hall, Inc., 1961). The various viewpoints in the Turnerian controversy are competently handled in George Rogers Taylor, ed., *The Turner Thesis*, Problems of American Civilization, Vol. 3 (Lexington, Mass.: D. C. Heath & Co., 1956); and Richard Hofstadter and Seymour Martin Lipset, eds., *Turner and the Sociology of the Frontier* (New York: Basic Books, Inc., 1968).

The contention that the general attributes of nation and citizen originated in the frontier or in its availability as a "safety valve" for an overcrowded East cannot be supported by factual analysis or general cultural theory. Given its very thinly occupied territory, with minimal social or political coherence, a precarious economy, and quite defective communications both internally and with the well-settled portions of the nation, it is hard to visualize any mechanism whereby the frontier could have wielded a commanding influence on the evolving life of the United States. Since the percentage of national territory or population accounted for by the frontier was slight at any given time (except during the earliest years of European settlement), this would have been a case of the tail wagging the dog. Furthermore, all that we know of cultural process indicates that innovations tend to diffuse outward from nodes of relatively intense social and cultural interaction, usually from those core areas where there is a concentration of power, wealth, and talent. Only in rare instances do new ideas or inventions seep back from periphery to nucleus.

This argument may seem to run counter to our earlier espousal of the Doctrine of First Effective Settlement, but on closer examination it does not. The identity and behavior of the first effective settlers were indeed of critical importance for the later development of their own areas, in both the colonial Atlantic Seaboard *and* derivative frontier zones. The Turnerian thesis, on the other hand, implies that despite the variety of peoples, economies, and other aspects of frontier life, there was some transcendent set of general conditions—an overall "frontierness"—uniting all segments and epochs of the American frontier. This overriding quality, in turn, ultimately shaped the personalities and behavior of all our citizens and helped mold the nation's fundamental institutions. That is a proposition that has not been established through solid historical research and probably cannot be.

If the Turner thesis lacks a firm basis in historical data, we are still left with the phenomenon of its remarkable vitality in academic circles and also, in less articulate fashion, in popular thought. Responding to this fact, one can suggest that if the various settlement frontiers failed to act upon the larger society and culture in ways we can measure, nonetheless the *idea* of a potent frontier spirit has been very much alive in the general consciousness. Mythological fact may be as triumphant as historical fact. Seen in this light, Turner's writings emerge as the first conscious evocation of a great quasi-religious folk legend, a codification and celebration of a fundamental article of the national ethos. It remains to be seen why Americans, most of whose ancestors had no direct encounter whatsoever with frontier life, should reach out almost instinctively toward the rather mystical notion that the unique strengths and virtues of the country are the legacy of a heroic frontier era. But the fact of such a groping is incontestable; and since nothing is stronger or more durable than an idea, it is almost irrelevant whether it is connected with actual events and conditions or not. To the degree that the Turner thesis is a formal declaration of an important state of mind, putting aside as secondary the question of strict historical veracity, it is valid and useful. But if we wish to discover how American culture really developed, we must turn elsewhere.

Why Did A National Culture Come to Be?

Thus far we have managed to evade perhaps the most fundamental question that could be asked about American culture: Why does it exist as a distinct national entity, with a certain coherence and personality of its own and with a recognizable continuity over so broad a territory and within so great a population? For there is no doubt that such a basic cultural and psychological unity does prevail, beneath the much more visible monolithic economic and political entity. The processes whereby the transplanted cultures of modern North America came to diverge from those of the ancestral strains of the Old World have been sketched. The relative contributions of various areas and of numerous national groups arriving in varied number at different periods have been described. But still it is not clear why a sense of budding nationhood should have surfaced among the people of thirteen British North American colonies (but not those in Canada or the West Indies) in the latter half of the eighteenth century, or even less whether it was inevitable, given the centrifugal forces of great distance, disparate habitats and ethnic and religious characteristics, and varied social, political, and economic interests.[33] A balkanized North America would appear to have been the more likely alternative. (Consider the fragmentation of New Spain after 1820.) Then there is the further difficulty of accounting for the survival, growth, and spread of the new nation and its freshly minted cultural pattern, despite formidable odds. Obviously, there was a complicated set of factors—geopolitical and commercial prominent among them—interacting in very complicated ways to bring about the birth of the United States. It is equally clear that once the viability of the political state has been explained, its cultural cohesion readily follows, or that this sequence of steps could just as easily be reversed.

Fortunately, such an explanation, although highly desirable, is not truly central to our purpose. Furthermore, it is simply unavailable as yet, except as a series of unproved speculations. Let us, then, accept as a primitive postulate the improbable fact of the existence of this large, populous, apparently durable nation-state with its special cultural geography. We turn next to an examination of what manner of special identity was wrought from the eclectic cargo of people and ideas carried to these shores from the Old World.

---

[33] For some stimulating approaches to this question, consult Karl W. Deutsch, "The Growth of Nations: Some Recurrent Patterns of Political and Social Integration," *World Politics*, 5 (1952), 168–95; Hans Kohn, *American Nationalism; an Interpretative Essay* (New York: Crowell Collier and Macmillan, 1957), and Richard L. Merritt, *Symbols of American Community, 1735–1775* (New Haven: Yale University Press, 1966).

CHAPTER TWO

Identity

On Capturing the American National Character

Anyone seeking to explore the special qualities of American culture finds a Comstock Lode of source material. The young Republic and its denizens were objects of intense curiosity on the part of visitors from the Old World, many of whom were not at all reticent about putting their observations into print; and the popularity of "America-watching" has slackened only slightly in recent years.[1] In addition, Americans themselves have been deeply concerned, almost obsessed, with their image and with their public and private identity, so that national self-analysis has been a thriving indoor sport among a number of native writers. In fact, this narcissistic attitude is itself one of the more extraordinary facets of the national character. Thus the literature dwelling on the peculiar nature of the American scene would fill many yards of shelf space; and much of the better material is of direct value to the cultural geographer. Some of the more precious insights into the basic design of American culture are to be derived not only from the better travel accounts and scholarly probes but from the revelations of creative artists. Such authors as Emerson, Thoreau, Whitman, the quintessential Mark Twain of

[1] Comprehensive guides to the early travel literature for the United States are to be found in E. G. Cox, *A Reference Guide to the Literature of Travel,* Vol. 2, *The New World,* University of Washington Publications in Language and Literature, Vol. 10 (Seattle: University of Washington Press; 1938) and W. G. Vail, *The Voice of the Old Frontier* (Philadelphia: University of Pennsylvania Press, 1949). The most important group of foreign commentators are listed and analyzed in Jane Louise Mesick, *The English Traveller in America, 1785–1835,* Columbia University Studies in English and Comparative Literature (New York: Columbia University Press, 1922) and Max Berger, *The British Traveller in America, 1836–1860,* Columbia University Studies in History, Economics, and Public Law, No. 502 (New York: Columbia University Press, 1943).

"Huckleberry Finn," Henry James, Hart Crane, Robinson Jeffers, Eugene O'Neill, and F. Scott Fitzgerald speak to the central issues, and so too the music of Charles Ives, George Gershwin, or Aaron Copland, the choreography of Martha Graham, and the architecture of Frank Lloyd Wright. More oblique and gnomic, since they purport to deal with extra-American realms of experience, but perhaps ultimately even more illuminating, are the works of Herman Melville or Austin Tappan Wright.[2]

In formal scholarly analysis, two fields of research have actively worked to delineate the core of American culture: American Studies and community studies.[3] Outside these two academic groups, several outstanding individuals, mainly in the fields of history or higher journalism—F. J. Turner, Alexis de Toqueville, Walter Prescott Webb, T. J. Wertenbaker, Perry Miller, Oscar Handlin, and Daniel Boorstin, for example—have delved into the mysteries of the cultural identity of the United States with noteworthy results. Among geographers, there has been a modest amount of work on specific cultural phenomena (see chapter 3), but surprisingly little aimed at questions of essential theme or process. Indeed the only explicit thrust in that direction would seem to be the work of J. K. Wright and David Lowenthal and of French geographers André Siegfried and Jean Gottmann.

[2] The reference is to Austin Tappan Wright, *Islandia* (New York: Farrar and Rinehart, 1942), an extraordinary novel about an early twentieth-century American's encounter with a vividly imagined continent, a volume whose great literary and extra-literary merits are, like those of *Moby Dick*, only slowly and posthumously coming to be recognized. For a brief, but warm, appreciation, see Lawrence Clark Powell, *A Passion for Books* (Cleveland: The World Publishing Company, 1959), pp. 97–111.

[3] The interdisciplinary entity known as American Studies, which is scarcely three decades old, has displayed much vigor, despite uncertainty over methodology and larger objectives. It represents the efforts of a cluster of scholars in the humanities, chiefly American literature but also history, music, architecture, folklore, and art, and some sociologists and anthropologists as well, to combine the techniques of cultural anthropology and humanistic analysis in interpreting American civilization holistically, and possibly contributing something toward a general theory of cultural process and change. American Studies is analagous to a number of other regionally oriented interdisciplinary consortia, for example, Sinology, Indian Studies, or Islamic Studies; but one may detect the implicit assumption that American civilization may differ in kind from others, that the central questions are unusual in nature and urgency. This movement has produced a number of works of interest to the geographer, and its practitioners merit close attention on the part of anyone seeking to define the essence of America. The topic is comprehensively and critically reviewed in Robert Meredith, ed., *American Studies: Essays on Theory and Method* (Columbus, Ohio: Charles E. Merrill Books, Inc., 1968).

An impressive series of community studies has been executed by a number of sociologists and anthropologists. In analyzing the social structure and behavior of small groups, usually within a single city or other small locality, these studies have confronted cultural issues, but generally as items of secondary concern. Their contribution toward the charting of national cultural patterns has been implicit rather than overt, but still of substantial value. The accomplishments of the community studies movement are summed up in Maurice R. Stein, *The Eclipse of Community; an Interpretation of American Studies* (Princeton, N.J.: Princeton University Press, 1960) and Conrad M. Arensberg and Solon T. Kimball, *Culture and Community* (New York: Harcourt Brace Jovanovich, Inc., 1965).

Much time and energy has been squandered by social scientists in debating the validity of the concept of "national character."[4] There has been much confusion over the meaning and implications of the term, and even more over the wisdom of applying generalizations about a nation to the behavioral patterns of its individual citizens. There is also legitimate fear lest studies in national character perpetuate—or create—false national stereotypes. Fortunately, for our purposes, this Gordian knot can be severed neatly with three observations:

1. Useful nonstereotypic statements can be made about the cultural idiosyncracies (that is, national character) of an ethnic group taken as a whole;
2. the population of the United States does indeed form a single large, discrete ethnic group;
3. statements about the character of the larger community cannot be, indeed should not be, transferred to individuals because of sharp discontinuities of scale.

These points deserve some elaboration.

The validity of national character analysis is dubious when a political state is an arbitrary slice of a larger national entity or is inhabited by a heterogenous collection of unassimilated peoples. What useful general statements could one make about the national character of Cyprus, Malaysia, Togo, or Surinam? It is an utterly different story when nation-state and ethnic group coincide. By definition, ethnic groups form one of the principal categories of cultural communities. Along the cultural continuum from the individual or family to such vast, diffuse entities as the Sino-Japanese or the Greater European Cultural Realm, the ethnic group stands at a critical juncture. It is the largest group to arouse a sense of gripping emotional commitment, one that in moments of crisis may be more precious than life itself; its members share a common symbol pool and are co-owners of the same basic value system and historical or mythic memories. The true ethnic group is also a true nation, in the most primitive, meaningful sense of that term. The ethnic nation may find political consummation in a nation-state, or, if it is less fortunate, it will be a submerged fraction of a larger political entity or distributed over the territory of two or more such states. The origin of ethnicity is immaterial. The group may date back to prehistoric times, with traditions so deeply rooted in a specific tract of land that the emergence of a modern nation-state is a spontaneous, almost unconscious act. In some instances, such as Yugoslavia, Czechoslovakia, Israel, or India,

4 The anthropological literature on the topic, which is large and inconclusive, has been reviewed in P. J. Pelto, "Psychological Anthropology," in B. J. Siegel and A. R. Beals, eds., *Biennial Review of Anthropology 1967* (Stanford, Calif.: Stanford University Press, 1967), pp. 140–208; and J. J. Honigman, *Personality in Culture* (New York: Harper & Row, Publishers, 1967). An excellent general bibliography is available in Michael McGiffert, ed., *The Character of Americans; A Book of Readings* (Homewood, Ill.: Dorsey Press, 1967), pp. 364–68. The most recent explicitly geographic treatment of the question is Robert D. Campbell, "Personality as an Element of Regional Geography," *Annals of the Association of American Geographers*, 58, No. 4 (1968), 748–59.

there may be some real or legendary basis for affiliation; or the new ethnic group may be a totally modern assemblage from immigrant or conquered groups, as has been the case with Canada, Brazil, Argentina, or the Soviet Union.

It is suggested that the population of the United States is a genuine ethnic group recently fused from quite miscellaneous materials through a combination of spontaneous fermentation and deliberate engineering by a powerful socioeconomic state apparatus. This group very nearly coincides with the political territory governed by Washington; but one must exclude the more culturally recalcitrant aboriginal Indians, the small Gypsy group, and a few other tiny, aberrant, or transient groups, and the people of Puerto Rico, the Virgin Islands, Samoa, the Trust Territories of the Pacific, and other alien areas currently under American rule. Similarly, there are minor, mostly temporary, exclaves on foreign soil of persons who are totally American in identity.

Once again, by definition, every ethnic group, including the American, possesses a distinctive cultural pattern. We have already agreed that American culture is a member of the Greater European Cultural Realm and no other, even though it may have been affected to some slight degree by non-European cultures. Our immediate problem is to identify those attributes of American culture that are particularly distinctive and important from the viewpoint of the cultural geographer. A complete catalog of American culture would amount to an almost endless list; and even given perfect knowledge, compiling it would prove to be a Herculean chore.[5]

What we are after is the most critical set of cultural attributes, the small package that will most parsimoniously signal the complex totality. The term "culture core" perhaps states most succinctly what is desired. As originally used by its author Julian Steward,[6] it refers to the basic livelihood arrangements of a community, those aspects of the human ecosystem or man-land relationships whereby a group obtains those goods deemed necessary for an adequate existence. Although understanding this system is, of course, critical to the study of any culture, the American livelihood pattern is not basically different from that of several other advanced nations.

As an alternative definition of the culture core, consider instead the central assumptions, the basic values and axioms that define aspiration and direction, even if they seldom describe actual performance.[7] The

---

[5] Some of the more useful items are listed in this volume under *Selected References*, pp. 186–211.

[6] Julian Steward, *Theory of Culture Change* (Urbana: University of Illinois Press, 1955), p. 37.

[7] This point is adumbrated in a remarkable chapter, "The Influence of Manners and Religion on Democratic Institutions in the United States," in a remarkable book, Alexis de Toqueville, *Democracy in America* (New York: Oxford University Press, 1947), pp. 195–220. This was initially published in Paris, 1835–1840. The term "manners" is never adequately defined, but in context appears equivalent to "ethos" or "basic values." De Toqueville states his argument vigorously: "The importance of manners is a common truth to which study and experience incessantly direct our attention. It may be regarded as a central point in the range of human

ethos—that elusive but powerful mood of this and all other distinctive cultures—is the product of many forces. Probably all facets of human experience enter into it, and it, in turn, molds the universe of its participants and the ways they comport themselves within it. Ethos changes through time, but slowly, so that there is a recognizable continuity between the mental world and shared values of colonial Americans and those of our contemporaries. There are also notable variations from place to place, but well within the limits of a sharply bounded national cultural system.

Unfortunately, there is as yet no rigorous, objective method for arriving at the nonmaterial culture core of an ethnic group. The analyst is obliged to feel out a hazardous subjective route. If the following scheme—admittedly an experimental one—is only one of a number of possible ways of defining the central attributes of the national character, it does have the advantage of being congruent with the judgments of many perceptive observers of the American scene, and of being supported rather well by those aspects of the land and people that can be mapped or measured.

Given all the foregoing conditions, it seems possible to chart the geographically relevant, locally distinctive character—the ethos—of American culture in terms of four pervasive themes or motifs, as follows:

1. an intense, almost anarchistic individualism;
2. a high valuation placed upon mobility and change;
3. a mechanistic vision of the world;
4. a messianic perfectionism.

The importance of these motifs, in both absolute and relative terms, varies markedly within the country and as between historical periods, but not so much as to vitiate their validity. This strategy of characterizing so vast and populous a nation in so very curt and cavalier a fashion would not be workable were there not a surprising degree of relative uniformity among the various regions and social segments of the country. As R. M. Huber describes it, "American culture, by its astonishing standardization of thought and uniformity of values in relation to size and population, permits us to make generalizations which would be far more difficult—perhaps impossible—to formulate for many other countries in Western Civilization."[8]

Keep in mind again that we are describing a culture, not the individuals who participate in it. Obviously, a culture cannot exist without bodies and minds to flesh it out; but culture is also something both of *and beyond* the participating members. Its totality is palpably greater

---

observation, and the common termination of all inquiry. So seriously do I insist upon this head, that if I have hitherto failed in making the reader feel the important influence which I attribute to the practical experience, the habits, the opinion, in short, to the manners of the Americans, upon the maintenance of their institutions, I have failed in the principal object of my work." (p. 213)

8 R. M. Huber in Meredith, *American Studies.*

than the sum of its parts, for it is superorganic and supraindividual in nature, an entity with a structure, set of processes, and momentum of its own, though clearly not untouched by historical events and socio-economic conditions. The example of language is relevant, for it is both a major component of, and paradigm for, the totality of culture. Languages have a life of their own, and the forces that change their content and structure are beyond the conscious control of any set of individuals. Furthermore, the basic syntax of language, its axiomatic implications concerning the meaning of meaning, are a decisive factor in the ways we think about and look at the world. It is just as correct to say that a language speaks us as to declare that we speak a language. In the same way, a culture practices us even more than we practice it. Thus the fact that most American citizens are not stereotypical embodiments (but how many are!) of the four postulates briefly noted above does not destroy their usefulness as illustrations of the essence of American culture. (It is a moot point, and one falling outside the orbit of this essay, whether it is possible for participants in a culture to alter its basic structure, even if they become fully, unhappily aware of it.)

None of the critical themes treated here is uniquely American. In fact, at least three may be found in similar form in other places or periods. But the particular strength and conformation of each and, most especially, their specific combinations are emphatically unique. Moreover, these traits are all borrowed from European sources in which they were latent, and they reflect a continuing interaction with Europe. If the special American tangent is to some degree idiosyncratic, we also have here a behavioral pattern that fits beautifully the needs of the modernization process, unquestionably the major event of recent world history. In fact, it would be difficult to imagine a culture more efficiently programmed for expediting progress toward the most advanced technologies and socioeconomic organization; and for at least the past 70 years the United States has indeed held a commanding lead in most phases of modernization. Are we witnessing a cultural pattern responding to the inner logic of the modernization process? Or has this process been initiated (by cognate cultures elsewhere earlier) and then been catapulted upward by the extraordinary favorable thrust of American culture? It is probable that both situations have prevailed. In any event, if there is no necessary association between the specific American cultural pattern and the phenomenon of modernization, in all the countries where it has occurred there are some interesting resemblances.

The Theme of Individualism in American Culture

The heroic self-image of the lone, self-reliant, upward-striving individual, sharing equal rights and opportunities with all and liberated from the shackles of tradition and authority, is possibly the single most dominant value in the cultural cosmos of the American. This fanatical worship of extreme individualism, indeed an almost anarchistic privatism, affects so many phases of our existence so deeply that no one can interpret either the geography or the history of the nation sensibly without coming to grips with it. To a much greater degree than the other three themes treated below, this idea is something of which most citizens are actively, even fiercely, aware; it flourishes at the personal as well as the collective level. This creed reflects an ideal goal rather than an everyday reality. Few persons have the resources to carry individualism to such methodically logical extremes as did, say, Henry Thoreau or William Randolph Hearst. But, nonetheless, the reverence for individualism has been a critical force in so many decisions, both private and collective, that it has had an incalculably strong effect in shaping the cultural landscape.

Americans are so conditioned to accepting individualism as the only proper or natural mode of behavior that it is difficult for them to realize that it is actually a quite late, deviant development in social arrangements. Until a few centuries ago, everyone's behavior was tradition-bound. The rules for personal, social, economic, political, and all other forms of activity were a matter of inherited or ascribed status in a family, sex, caste, village, occupation, or church, which, in turn, was determined by the accidents of place of birth and ancestry. It was only with the dawn of the modern age in Northwest Europe, and later elsewhere, that individualism became either possible or widely accepted. In fact, one can now neatly distinguish the modernized, relatively advanced and prosperous countries of the world from the traditional, less advanced societies by the presence or absence of the individualistic philosophy. And any recent success in achieving modernization can be measured by noting how firmly individualism has taken root in the evolving country. What is truly remarkable in American culture is not the existence of a trait shared with the people of Sweden, Switzerland, Rhodesia, Hong Kong, and Siberia, but rather the extremes to which it has been carried both ideologically and materially.

The doctrine of individualism has expressed itself in many ways, but there are three particularly interesting forms: the Frontier Myth; the Protestant Ethic; and the Success Ethic. The apotheosis of the frontiersman as the resourceful, isolated fighter against the wilderness, triumphantly carving out his own autonomous barony, the virile libertarian, jack-of-all-trades, and rough-and-ready paragon of all democratic virtues, has captured the popular imagination, and has survived lustily in folklore, drama, and fiction. To a surprising degree, this is still the role in which many Americans cast themselves in their private daydreams. The result has been a search for the physical or moral equivalent—in technology, the arts and sciences, overseas, in outer space, and elsewhere—for the vanished settlement frontier. There is also an atavistic acting out of the Frontier Myth in many phases of the settlement landscape, recreation, and political behavior.

The familiar, if controversial, Protestant Ethic has, it is claimed, been a potent force in American economics and theology,[9] though quite

9 The thesis of the Protestant Ethic was initially presented in Max Weber, *The Protestant Ethic and the Spirit of Capitalism* (London: G. Allen & Unwin,

unevenly as between different epochs and regions. The emphasis laid upon individual spiritual salvation seems to have been translated into material terms as the parallel doctrine of worldly salvation through constant industry and frugality. Predictably, its self-denying practitioners took a rather dim view of aesthetic experience. This and their hostility toward idleness and pleasure for its own sake have lingered on in the American mind, and have not been completely erased even today.

The Success Ethic, one which is dominant in the United States today, though hardly unique to this country, seems to be a rather simplified product of the Protestant Ethic. Instead of spiritual salvation, however, it holds the greatest good to be the greatest possible individual success. Unfortunately, the definition of success and the scales for measuring it are not clearly understood. Some look upon it as a form of inner satisfaction or ego gratification, but most see it competitively, as the attainment of maximum social approbation, envy, or status, or power. However, under the fluid social conditions prevailing in America, status and power are elusive, and absolute personal success is seldom certain. Contrary to popular myth, simple monetary greed is not the governing passion in American personal behavior, as an expression of individualism or anything else. The bona fide miser is a newsworthy rarity in the United States. Money is a neat numerical device for keeping the score in the scramble for success, not the goal itself. Conspicuous generosity is a mark of success; and the American who has arrived may work as hard at giving away or spending his wealth as he did in piling it up.

Beyond any question, the unusual freedom and mobility given the individual to capitalize upon his innate capacities and to create or grasp new opportunities has been as potent a factor as any in the unequalled material accomplishments of the Americans. But the other side of the coin is the heavy psychological price they pay for freedom and prosperity. It is found in a constant sense of insecurity, of having no final anchorage in things, men, or the supernatural. The American is born into an uncertain place in an unstable society, never quite sure of his identity, and spends his life in a quest for an elusive ego satisfaction. The powerful solvent of individualism has begun to weaken even the most intimate and cohesive of social groups, the nuclear family.[10] The

---

1930) and is critically discussed in Robert W. Green, ed., *Protestantism and Capitalism; the Weber Thesis and its Critics* (Boston: D.C. Heath and Company, 1959). The complex interplay among theology, business practice, and science in early New England, an area of peculiar significance for the later evolution of the nation, is treated in Dirk J. Struik, *The Origins of American Science (New England)*, rev. ed. (New York: Cameron Associates, 1957).

[10] The point has been made by a Chinese-American anthropologist with the unusual advantage of being able to view American culture from both inside and outside. "We have seen that mutual dependence is the outstanding Chinese characteristic, and that this deep-seated tendency to rely upon other persons, especially those within the primary group, produces in the Chinese a sense of social and psychological security. . . . The self-reliant American, however, strives to eliminate from his life both the fact and the sense of reliance upon others. This unending struggle to be fully independent raises the threat of perpetual social and psychological insecurity. A close parental bond is severed early in life; the marriage which replaces

consequences may be unpleasant for both individual and society. Furthermore, the practice of individualism has sometimes gone past the point of optimum return and has become dysfunctional, even in a narrow material sense.

The enthusiastic pursuit of individualism has yielded some paradoxes. Americans are at the same time the most acquisitive and the most generous of folk; they are intensely private people, yet at the same time unusually gregarious; and individualism goes hand in hand with sheep-like conformity. These seeming contradictions do make sense, however. The accumulation and the giving away of property are equally proud badges of success, the one following the other. The rootlessness of the private person, lacking any secure base in family, clan, neighborhood, or any other system of mutual dependence, impels him to turn to a series of churches, political causes, lodges, clubs, jobs, and other shallow, transient associations. Masses of insecure individuals find refuge and solace in mob camaraderie, in mass conformity within narrowly prescribed ranges of individualistic behavior.[11] (The genuine certified eccentric abounds in Great Britain, where the individualist can indulge his whims as a kind of obligato above a relatively steady social base; he is almost unknown in the United States.) In the final analysis, individualism, as it is enshrined in national tradition, is hardly a viable way of life in a complex, advanced society of 205 million persons. The illusion of its supremacy and potency may fade in some sectors, notably the economic, but the idea survives. Faith in the absolute goodness of letting every individual do almost anything he wishes is writ large across the American landscape. It will be a long time before the student of American cultural geography can ignore its pervasive effects.

Individualism Manifested in the Cultural Landscape

POLITICAL PATTERNS.   If the individualistic strain in the American cultural fabric reveals itself in many geographic phenomena, it does so as strikingly and basically in political behavior as in any other form.

---

it is unstable; heroes come and go; class affiliation is subject to a constant struggle to climb from one level to another; and the alliance with God, though less ephemeral than the preceding relationships, is nevertheless affected by the same divisive forces. For, given the ideal of total self-reliance and its concomitant version of a God who helps those who help themselves, Americans seek their final anchorage in harbors other than men or the supernatural." Francis L. K. Hsu, *Americans and Chinese: Two Ways of Life* (New York: Abelard Schuman, Ltd., 1953), pp. 279–80.

11 And such mass conformity can hardly produce any genuine sense of community. It is not putting things too strongly to state: "There is very little true local community or government in the United States. We are a collection of shallow and spurious and artificial communities, and it is one and the same thing to say about us that we need more national government and that we have hardly any real local government at all. . . . What we do have—mostly the corporations and similar 'private' organizations—is barely recognized for what it is." David T. Bazelon, *Power in America: the Politics of the New Class* (New York: New American Library of World Literature, Inc., 1967), p. 44.

Imbued with the notion of the sovereignty of the individual, the American generally regards government as a necessary nuisance at best and thinks that government best which governs least. The overtones of the epithet "politician" have always been unflattering. Fear and suspicion of authority run deep, and the more remote the seat of power, the sharper the antagonism toward it. The fiction that each of the 50 states is indeed a sovereign kingdom is one that many in public life actually seem to believe, and to which many more pay lip service. Many a function that could be performed more efficiently at less cost at the federal or regional level is jealously hoarded by the state, county, or even municipality.[12] It is truly remarkable—and contrary to much expert early opinion—that the United States achieved federal union in the 1770's and has maintained it continuously ever since. This fact argues the existence of other forces, and, most basically, that mysterious sense of nationhood combined with the free migrations of many persons within the national space, powerful enough to counteract the intrinsic anarchism of the American. But if Americans by the millions were prepared to lay down their lives for the abstract idea of Union, few would lift a finger to strengthen the national government. Indeed many pious platitudes are still being uttered about the urgency of decentralizing administrative functions now based in Washington when, in fact, the national government is still woefully impotent in many vital areas. Among the larger advanced nations of today, one would have to turn to other neo-European lands of similar bent, for example, Australia or Canada, to find local affairs so feebly managed or even ignored by the central authorities. There is something deeply symptomatic in the absence of a national police system[13]—the very thought of it causes most Americans to recoil in horror—or the parcellation of the National Guard into 50 quasi-sovereign entities. Even in so simple a matter as a national system of highway signs, Americans find it impossible to reach consensus.

The chronic antiauthoritarianism of the people has made it virtually impossible for governmental agencies to chart or carry out any form of long-range social and economic planning, except fitfully during such cataclysms as World Wars I and II or the Great Depression. Indeed it was not too long ago that the term "planning" was considered a dirty word by many Americans. Effective planning is practiced, for quite valid, selfish reasons of profit and efficiency, only by the larger privately

[12] As of 1967, there were no less than 3,049 counties, 18,048 municipalities, and 17,105 townships in official existence in the United States; and these local governments, in turn, called upon the services of 508,720 *elected* officials, plus many more appointed employees on wages or salary. U.S. Bureau of the Census, *Statistical Abstract of the United States: 1969*, 90th ed. (Washington, D.C.: Government Printing Office, 1969), 405. The amount of administrative inefficiency implicit in these figures staggers the imagination.

[13] Yet, in keeping with the ambivalence of the national character in matters of privacy, there is little general indignation over the elaborate surveillance systems being developed, often quite covertly, by various governmental bureaus and credit organizations, for the collection and filing of the most intimate kinds of information about individual citizens.

owned corporations and the military and technological branches of the government. The dynamics and larger design of American life, and thus of our human geography in general, are the products of autonomous, essentially uncontrolled social and economic forces and the interplay of numerous competitive special interests.[14]

In spatial terms, the fragmentation of political authority is extraordinary, and it is not just a product of historical happenstance. This is most painfully obvious in the urban centers where some three quarters of the population now lives or works. Except for some smaller places, cities, which operate as unified socioeconomic areas, are balkanized into a number of politically sovereign entities. The larger the metropolis the more numerous and irksome these divisions, and the greater the problems of effecting treaties and alliances whereby the urban area can somehow transact its business. In extreme cases, a single metropolitan area may contain several hundred distinct governments. There are not simply the nominal territorial units (city, village, township, and so forth) but also a multiplicity of special jurisdictions for administering water supply, sewage disposal, pollution control, highways, harbors, schools, hospitals, airports, police, planning, zoning, parks, charities, and many other things. Each of these has some degree of independence. Clearly the costs in time, money, and mismanagement are burdensome. Attempts to rationalize the situation through annexation of outlying places or by creating consolidated metropolitan governments have met with very modest success. Of the 230 Standard Metropolitan Statistical Areas now in existence, only one, Miami (Dade County), Florida, has anything approaching a unified government; and even that experiment is not yet beyond jeopardy. Superficially, the explanation for the slow progress of the "metro" idea lies in the technicalities of state law. More fundamentally, however, the crazy quilt of local authority reflects a deep-seated American individualism; laws and institutions ultimately express the inner temper of the collective mind, in this case a centrifugal, anarchistic mood, the displaced frontiersman's deep, dark dread of "city hall." Psychically, the stockade around the suburban fort seems to be worth the heavy penalty.

In rural areas, the persistence of some three thousand county governments and tens of thousands of minor civil divisions into the late twentieth century cannot be explained in rational terms. These miniature principalities did make sense in the days of horse-drawn vehicles or even the railroad, but are completely anachronistic in an automotive era. In any

---

14 One of the more remarkable facts about a society that is normally so engrossed with facts and figures is that so very little is known about who owns the physical and productive resources of the nation and how basic decisions are reached concerning their allocation, or other matters of national life or death. It is also remarkable that so few academics have bestirred themselves to find out. Apparently this is the most indelicate of subjects, much more so than sex or any of the standard vices. For a massive, irascible, fact-laden, and ultimately persuasive attempt to explore the unmentionable, see Ferdinand Lundberg, *The Rich and the Super-Rich: a Study in the Power of Money Today* (New York: Lyle Stuart, 1968).

case, the majority of these areas have suffered absolute loss of population. Only a single state, Connecticut (the location may be significant), has stripped its counties of obsolete functions. Once again, the symbolic value of these vestiges of a bygone democratic rurality may operate, if only at the subconscious level, to prolong their life.

SETTLEMENT PATTERN.   Nowhere is the individualism that suffuses the American cultural pattern so legibly inscribed as in our rural settlement patterns or in that general ruralism that is inescapable in American thought and action. The isolated rural farmstead readily became the dominant form of settlement throughout British North America, just as the single-family dwelling is now standard in all but the oldest cities. In the case of New England, recent research has established the fact that, with rare exceptions, settlement was dispersed in form—contrary to popular myth.[15] Other colonial projects for farm villages, for example, the plans framed by William Penn for the immediate Philadelphia hinterland, quickly came to naught, if they were actually implemented at all. Except for the firmly disciplined Mormons and a few other pietistic groups, all subsequent attempts to cluster farm families met with failure.

The explanation for the prevalence of the isolated farmer lies neither in European precedent nor in economic logic. Many of the early immigrants who were destined for American farms had probably been city-dwellers. Although the European source areas were among the most highly urbanized anywhere in the seventeenth and eighteenth centuries, by conservative estimate more than 90 percent of colonial Americans were rural residents.[16] Most of the colonists drawn from rural portions of Europe must have been living in clustered farm villages, since that mode of settlement was general there. Where it was not, the distance to a central place and the rural dispersion of traditionally urban or village functions tended to be less than in the New World.

After 1787, with remorseless rectangularity and with the greatest possible disregard for the sphericity of the earth and the variable qualities of its surface, more than two million square miles of public domain were blocked off into 36-square-mile townships and these, in turn, subdivided into 36 sections, each one mile square. This simplest of possible geometries represents a distillation of earlier Colonial experience and thought, especially that of Virginia and New England—an indefinite projection westward of the Jeffersonian model of an agrarian republic

[15] Glenn T. Trewartha, "Types of Rural Settlement in Colonial America," *Geographical Review*, 36, No. 4 (1946), 568–96; Joseph S. Wood, "Village and Community in Early Colonial New England," *Journal of Historical Geography* 8 (1982), 333–46.

[16] For a detailed account of the most rural of the American colonies, see Harry Roy Merrens, *Colonial North Carolina in the Eighteenth Century: A Study in Historical Geography* (Chapel Hill: University of North Carolina Press, 1966), esp. pp. 142–72.

peopled by sturdy independent yeomen. Note the lack of provision for town sites; any urban development was to be an excrescence upon a thoroughly rural design.[17]

The layout of roads and, more particularly, the shape of individual properties and the placement of farmsteads within them, reveal a strong bias toward solitude. It would have been feasible to combine rural residence on sizeable farms with a degree of sociability, as has been done in rural Quebec, by letting the farms run back a full mile or more from relatively narrow road frontages, so that homes could be set in close proximity.[18] Instead, individual properties tend to be compact or square; only exceptionally do three or four farmsteads occupy the common corners; and quite often the farmhouse is located so far back from the highway as to be completely hidden from its nearest neighbors. Although there are obvious functional advantages in living on one's farm instead of commuting daily from a central place, there are also social and economic drawbacks. A far more efficient road system could easily have been devised to serve at less cost an equal number of farmers holding the same amount of land, but with a great reduction in mean distance between homes.

As many ordinary central place functions as possible were ruralized, and some remain so to this day. Only in rare instances could the town be dispensed with totally, as happened in much of Tidewater Virginia or among the larger southern plantations.[19] Itinerant merchants were numerous until recently, the mobile library flourishes, and many completely rural county courthouses still exist.[20] In accordance with the Northwest Ordinance, rural schools prevailed, most of them no more than a single room in size. Rural churches seem greatly to have outnumbered urban congregations; and that thoroughly American phenomenon, the outdoor camp meeting, abounded in various times and

17 Excellent presentations of the historical geography of American systems of land division are available in Francis J. Marschner, *Land Use and Its Patterns in the U.S.*, U.S.D.A. Agriculture Handbook No. 153 (Washington, D.C.: Government Printing Office, 1959) and Norman J. W. Thrower, *Original Survey and Land Subdivision: A Comparative Study of the Form and Effect of Contrasting Cadastral Surveys*, Monograph Series of the Association of American Geographers, Vol. 4 (Chicago: Rand McNally & Company, 1966).

18 As described in Richard Colebrook Harris, *The Seigneurial System in Early Canada: A Geographical Study* (Madison and Quebec: University of Wisconsin Press and Les Presses de l'Université Laval, 1966); or R. Louis Gentilcore, "Vincennes and French Settlement in the Old Northwest," *Annals of the Association of American Geographers*, 47, No. 3 (1957), 285–97.

19 The impulse toward the spatial segregation of homes as an expression of individualism may well have been in the air in Europe at the time the United States was settled; but there were the practical difficulties of a landscape thoroughly encumbered with relict patterns and structures. When reform of the settlement pattern was effected, mainly for economic motives, as happened during the British Enclosure Movement or the parallel breakup of the agglomerated village in eighteenth-century Sweden, the practical problems were staggering.

20 Edward T. Price, "The Central Courthouse Square in the American County Seat," *Geographical Review*, Vol. 57, No. 1 (1968), 38.

places. These isolated religious establishments and the later grange halls and annual county fairs offered the physical basis for much of the organized social life of the nation. At the same time, there was a corresponding atrophy of the social side of the small towns, which, like the larger cities, have always been overwhelmingly economic in function. The recent dwindling away of rural schools and churches is more the result of declining farm population and improved transport facilities than of any slackening in the psychological impulses that led to their generation. Possibly the one visible landscape item that most poignantly discloses the intense individualism and rurality of the American is the isolated country graveyard. In sharp contrast to most European and early colonial custom, whereby burial was restricted to municipal cemeteries or to village or city churchyard, the American custom has been the family plot and bucolic community graveyard.

The contention here is that economic necessity was buttressed by cultural predisposition to make rurality more than a temporary measure: it became the pure, natural fruition of a powerful individualism. The evidence for this is the tenacious grip of rural values upon the mass mind long after the nation has ceased to be essentially agrarian, as can be shown in analyses of modern literature and political behavior. The somewhat fictitious self-sufficient family farm remains one of the durable central symbols around which American mythology is organized, for city folk as for others. Conversely, there lurks in the American mind a clear repugnance toward and distrust of the large city.[21] It is only against the background of such an antiurban bias that one can rationally explain the inordinately long delay in electoral redistricting of state legislatures so as to give urban voters equal representation. Likewise this bias explains that haunting suspicion in the city-dweller himself that he is a moral pariah and hence the political inferior of his uncorrupted country cousin. Similarly, the extremely rapid territorial explosion of the American city, faster and more patternless perhaps than that of any other nation, is undoubtedly abetted by the same sentiments. The diffuse urbanoid periphery represents an uneasy compromise between a rural tropism and the functional imperative of the city. The case for the significance of individualism in the American culture complex need not be made exclusively from rural evidence. The haphazard morphology of the American city and the nearly universal lack of any aesthetic or functional association between adjacent structures also speak volumes.

[21] This rural tropism and antiurban bias is rather overstated in Morton and Lucia White, *The Intellectual Versus the City* (Cambridge, Mass.: Harvard University Press and M.I.T. Press, 1962). The question is perceptively treated within a broader context in such works as Leo Marx, *The Machine in the Garden: Technology and the Pastoral Ideal in America* (New York: Oxford University Press, 1964); and Frank R. Kramer, *Voices in the Valley: Mythmaking and Folk Belief in the Shaping of the Middle West* (Madison: University of Wisconsin Press, 1964); while the theme of an ambiguous attitude toward cities is explored at length in Anselm Leonard Strauss, *Images of the American City* (New York: The Free Press, 1961). An adequate full-scale study of how Americans think and feel about cities, concretely or in the abstract, still remains to be done.

ECONOMIC PATTERNS. The theme of individualism in American economic activity is so thoroughly shopworn it scarcely calls for elaborate argument or documentation. Such catch phrases as "rugged individualism" or "the free enterprise system" eloquently testify to the American businessman's image of himself and the system to which he belongs. If there is a good deal of substance to the myth of the completely unfettered entrepreneur—and trustbusting would not have been socially acceptable were it not in basic accord with the theme of individualism—there is still a wide gap between the ideal and the real. American businessmen have indeed been a remarkably free-wheeling lot in comparison with their confrères in almost any other society, so that the speculator in land and commodities is not only tolerated but is an honored member of the community; but even the most rapacious "robber baron" was subject to a few restraints. The failure rate among smaller entrepreneurs has been dismal; and the majority of American farmers, now as in the past, have been either tenants or deeply indebted to mortgage lenders. The compelling logic of maximizing profit and minimizing risk has nurtured the giant, impersonal corporation; and entire industries find it most lucrative to eliminate serious competition and to regulate prices and other practices by mutual agreement. But reverence for the myth and an incurable antagonism toward truly effective government that is so integral a strain in the general cultural pattern make it impossible thus far to achieve a totally rational economy. Nowhere is this more evident, and nowhere are the geographic implications more direct, than in the transportation and communication industries.

To a degree unmatched in other lands, these industries have been controlled by (ostensibly) competitive enterprises owned by individuals or private corporations. This was true for earlier eras when some postal services, most improved roads, ferries, canals, and toll bridges—many of them needlessly duplicating the services of others—were in private hands. The history of the American railroad industry, understandable only against the backdrop of individualism, is one of chaotic growth, wholesale duplication of lines, and a frustrating proliferation of terminals in major cities. Only slowly, against a rearguard action by public opinion and federal bureaucracy, have the necessary, inevitable mergers taken place. Even now there is no truly nationwide line; and unlike almost every other modern nation, the coterminous United States has virtually no federally owned or operated trackage except a recently introduced, severely truncated passenger system. The recent history of air transportation is essentially a replay of the railroad saga, with the significant exception that technical factors have compelled the local consolidation of terminal facilities: the same duplication of lines and services; the same uneasy combination of competition and cooperation and grudging reliance upon governmental semi-control and subsidy. And what other self-respecting nation lacks that glittering status symbol, the national airline? The ghost of individualism still stalks the land.

At an even broader level, the universal mood of economic privatism has made it impossible to draw up, much less execute, the sort of comprehensive transport system that is urgently needed, one that would integrate railroads, trucking, airlines, bargelines, pipelines, busses, and mass transit in a manner that could save enormously in terms of costs, time, annoyance, and space. At the other extreme, the unrivaled supremacy of the private car in the ground transportation of passengers, and the correspondingly dismal results of efforts to maintain or revive mass transit systems, argues a deeply ingrained bias toward personal autonomy, symbolized by the metal-and-glass bubble around each American.

In communications, the picture is much the same. Only postal services are a federal near-monopoly; but perenially some business spokesman has uttered the primordial tribal cry of "Turn the Post Office over to sound private businessmen!" until the recent restructuring of the system within an autonomous quasi corporation. Again, in contrast to like facilities in every other civilized country, our telegraph, radio, and television facilities are, with minor exceptions, operated for private profit, and with wretched results. In the case of telephone services, elsewhere a government monopoly, the free enterprise system has acquitted itself more nobly, in part because a single corporation bestrides the field. But a glance at a map of franchise territories reveals a surprising number of local companies, and indicates that here again the theme of individualism has achieved spatial expression.

RELIGION AND EDUCATION. The stubborn insistence upon individual autonomy and the most absolute freedom of choice possible has yielded unique results in American religious life. Not only is the number of denominations and subdivisions thereof far in excess of anything recorded anywhere else at any time, but many groups united in name or theory are, in fact, almost totally disunited at the regional or congregational level (and dissension within the single congregation is a popular indoor sport). Diversity is not only doctrinal but also spatial and behavioral. The number and range of local churches is unequalled elsewhere, so that a rather small village may boast a half dozen denominations, and only rarely does a single creed dominate even a single locality. The American lacks any fixed locus in spiritual as well as in social or physical space. Members of a single family may belong to two or more churches simultaneously; a given individual may join several different churches in the course of a lifetime. He may even patronize two or more churches on successive sabbaths, in a thoroughly egalitarian spirit. In contrast to the economic and political realms, where it can be demonstrated objectively that individualism yields dysfunctional results in contemporary America, strong arguments can be adduced for either side of the issue of external freedom of choice in matters religious: it is a mixed blessing.

The structure of the American educational system also confirms the individualist streak in the national character. The prevailing pattern seems so natural to Americans that it is startling to realize how truly aberrant it is when set against worldwide norms. Public education at the primary and secondary levels—and often at the collegiate level—is under strict local control by municipality or county, with only a mild degree of supervision by the state government and none at all at the federal level. Although technical or financial assistance may be forthcoming from Washington, strenuous efforts are made to prevent overt interference with local school policy. Thus there is no federally operated system of schools, except on Indian reservations and in the military services, and any advocacy of such would be political suicide.

In addition to the proliferation of public, parochial, and private elementary and high schools found in many American communities, there is an abundance of privately endowed and operated colleges. Many are, or were, denominational, but most appear to have been founded for purposes of gain, philanthropy, the propagation of particular points of view, or the perpetuation of the donor's name. Private colleges are genuine rarities in other parts of the English-speaking world or elsewhere. Their profusion in the United States attests not only to the nation's material wealth but also to the free range given individual initiative. Little work has been done on the geography of education in the United States and none apparently on the historical geography of higher education. An analysis of colleges, especially the private liberal arts college, might yield important insights into the cultural geography of the nation.

MISCELLANEOUS SYMPTOMS. The extremely high valuation placed on individual effort and freedom in American culture surfaces in many other aspects of the national scene. The following items would seem to have considerable potential for the geographic analyst, but remain almost totally unexplored. The American hospital system offers a close parallel to our schools in its mixture of public, religious, and private endeavor and the occasional proliferation—and probable duplication— of facilities in certain fortunate localities. The emphasis on profit-making proprietary hospitals and homes for the convalescent and elderly is probably stronger than in other societies. Their spatial configuration awaits the explorer. The vast number of privately endowed philanthropies (about 18,000 at present), not only for schools and hospitals but for many other forms of activity as well, is the result, in part, of phenomenal affluence and the peculiar structure of the tax laws, but is also a ringing assertion, usually posthumous, of the personal triumph of the lavish individualist. The United States would also be far in the lead in any international competition in voluntary associations, whether in absolute quantity (more than 15,000 are listed in the standard source)[22] or number per capita.. These groups, which may be fraternal, athletic, recreational, charitable, occupational, business or professional, scientific, artistic, political, criminal, neighborhood, or related to any of a limitless variety of other special interests, reflect both a footloose individuality and the somewhat shallow gregariousness that develops among people adrift without a secure social

22 Margaret Fisk et al., eds., *Encyclopedia of Associations*, 2 vols., 6th ed. (Detroit: Gale Research Co., 1970).

base. Again, it would be most profitable to look into their geographic manifestations.

A final suggestion that could plunge one deeply into the American cultural psyche concerns the significance of attitudes toward firearms. There seems to be something peculiarly symptomatic of an extreme individualism in a rate of private rifle and revolver ownership far exceeding that of other urban-industrial lands. Clearly this is a socially dysfunctional phenomenon. Even more symptomatic is the sheer volume of the recent uproar over the real or imagined threat to what many consider an inalienable individual privilege. Once more, the seductive image of the frontier hovers over the land.

The Theme of Mobility and Change

All societies change. This truism has held since time immemorial, but it is only in the past few centuries that it has had any real meaning for individuals. Only recently have things moved so rapidly that one is unable to overlook major change within a single lifetime or shorter intervals.[23] For members of most cultures, change is profoundly disturbing and unwelcome, something to be resisted and shut out. The American attitude is remarkably different. Here necessity has become a virtue; the external imperative has been internalized. Once again we have a cultural trait pandemic throughout the modernized world, but with the United States blazing the way, riding the cab of the locomotive of history, happily stoking the fires. The love of change and of all forms of mobility, an innate restlessness, is one of the prime determinants of the structure of American national character. Living in a world of swift and sudden transitions and constant acceleration in the pace of innovation, Americans revel in the new and the rapid; they are sulky or bored with the old and the static. Choice, change, movement, and progress are the new gods. Nowhere else does the cult of youth have so many devotees; and nowhere else is it so uncomfortable or even disgraceful to reach an advanced age. It is no accident that the United States has spawned and developed jazz music to such sophisticated levels. Indigenous art forms reflect the core of a culture; and jazz ideally embodies the feeling of hurtling momentum, imbalance, fluid relations, and inner uncertainty.

Perhaps the briefest acceptable definition of the uniqueness of its culture is this: America is process. This understandably exaggerates a universal principle, for the comprehension of process is fundamental in studying any society; and process and structure are everywhere lashed together in a never-ending embrace. In the case of the United States, however, the processual side of reality is the more obvious. The rate and complexity of change are so overpowering and process so rapidly evolutionary that the resultant structural arrangements tend to be fleeting and

---

[23] The first book length discussion of this new phenomenon is Alvin Toffler, *Future Shock* (New York: Random House, Inc., 1970).

unobtrusive.[24] A more basic statement about the quintessential nature of its culture is that the United States is a country of constant decision. The existentialist point is well stated by Perry Miller:

> He who endeavors to fix the personality of the American in one eternal, unchangeable pattern not only understands nothing of how a personality is created, but comprehends little of how this nation has come along thus far... he fools himself if he supposes that the explanation for America is to be found in the conditions of America's existence rather than in the existence itself. A man is his decisions, and the great uniqueness of this nation is simply that here the record of conscious decision is more precise, more open and explicit than in most countries...being an American is not something inherited but something to be achieved.[25]

The ways in which time, space, and other apparently absolute elements of experience are actually perceived can vary widely from one cultural group to another.[26] One of the basic postulates of the American cultural system—and one to which nearly all its members subscribe—is that time is a peculiarly precious thing; literally, time is money, as we are told a dozen times a day. Another is the axiomatic belief in the directional, linear flow of time. In many, perhaps most, cultures, time is felt to be either patternless, with no discernible rhythm beyond that of the year or generation, and is seldom thought about consciously, or it is vaguely circular, with grand recurrent cycles and subcycles. In either case, events are enacted within an ultimately immutable universe. In the American cosmos, on the other hand, events flow purposefully and rapidly from a dim, unregarded past along a straight, narrow groove toward an indistinct but brighter future. This notion can be neatly symbolized as an arrow in upward trajectory. Such a linear patterning of time and temporal behavior, is, of course, closely paralleled by an equally linear patterning of space and spatial behavior.

As a group, Americans are profoundly insensitive to, or uninterested in, any segment of time except the present and the immediate future. Unlike more static cultures, whose members dwell companionably with memories or relics of past generations, the American is often nearly oblivious to his history, or even takes positive pleasure in demolishing older artifacts and replacing them with newer (= better) ones. What visible evidence is there, for example, that New York City has existed continuously for more than 340 years? Insofar as there is any enthusiasm for the past, it is for a neatly segregated, sanitized, romanticized package in museum form, vide Williamsburg, Greenfield Village, Sturbridge Vil-

---

24 The theme of incompleteness and rapid change in a future-oriented visible landscape is dwelled upon in David Lowenthal, "The American Scene," *Geographical Review*, 58, No. 1 (1968), 61–88.
25 Perry Miller, *Nature's Nation* (Cambridge, Mass.: Harvard University Press, 1967), p. 13.
26 Edward T. Hall, *The Hidden Dimension* (Garden City, N.Y.: Doubleday & Co., Inc., 1966).

lage, or the never-never land of the Western film—a synthetic past uncannily mirrored by the equally shallow, museum-display futures exhibited in our world fairs and the science fiction movie.[27] Although Americans are powerfully propelled toward a future in which present expectations will be consummated and current incongruities in our visible landscape are to be patched up, it is a very short-range future—the lifetime of living individuals. The remote future is almost never thought about.

The high esteem accorded time by Americans, who have a correspondingly weak awareness of space, is apparent in many ways. Geography courses are rare in secondary schools and lacking in many a college, but the teaching of history is well-nigh universal. (We may not have much true feeling for the past, but at least we feel we ought to.)[28] The number of clocks in public and private places is noteworthy. Most individuals feel almost naked without their wristwatches, but how many carry compasses, maps, or field glasses, or can identify cardinal directions or the local river basin? Here again the United States carries to an extreme a motif that first appears in early medieval Europe with the elaboration and first widespread, communal use of mechanical timepieces. Promptness is celebrated, and time is chopped into tiny slices to be doled out grudgingly. Particularly revealing is the relative status of time, space, and substance. Americans will go to great lengths to conserve time, including the abolition of distance for all practical purposes, and a spendthrift wastage of physical resources. In fact, almost anything is permitted if a few moments can be saved. Linear distances are automatically translated into temporal terms: it is so many hours between cities, or so many minutes between home and job. This interchangeable spatial-temporal linearity is strongly developed at the sacrifice of territorial acuity, so that the perception of two-, three-, or four-dimensional space is quite feeble in a land of ephemeral human landscapes.

The overwhelming emphasis on haste, speed, and change in the American cultural pattern has had endless repercussions upon the social and economic behavior of our citizens. An innate receptivity to innovation, a hunger for novelty, and a tolerance of, even an eagerness for, the built-in obsolescence of many products from paper tissues to skyscrapers have affected our social and economic geography in ways too numerous and obvious to be recounted here. Suffice it to say that even in the short run the immense, rapid turnover of things and people, through space and processual channels, cannot be fully justified on

[27] David Lowenthal, "The American Way of History," *Columbia University Forum* (Summer 1966), pp. 27–32.
[28] Books and periodicals dealing with American history in a popular style have proved quite lucrative in a number of instances; but, unlike Great Britain, Germany, and other countries, where such publications flourish, we have no successful mass-audience journal concerned with matters geographical. The financial success of some essentially touristic ventures or those dealing with natural history does not vitiate the argument.

strictly rational economic grounds or otherwise. Taking a deeper, longer view, it is clear that the vigor of the various American circulation systems —of migrant, artifact, information, and so forth—is less a response to narrow monetary impulse than to the essential mold of the cultural matrix. By hustling, wandering, experimenting, "developing" themselves and their land, and consuming space and treasure at a prodigious rate, Americans are scratching a very basic itch indeed.

In geographic terms, this inner restlessness is expressed most purely and simply as spatial mobility. The phenomenon can be classified into two major categories: *migration,* that is, change in permanent residence; and *circulation,* that is, temporary or pendular movements. Although the available data and measures for migration in the United States are far from ideal, they do clearly indicate that the great majority of Americans shift residence at one time or another and that a significant minority do so with great frequency and over considerable distances.[29] Unfortunately, the information at hand does not permit useful comparisons with other nations.[30] A survey taken in 1958 indicated that only 26 percent of Americans 18 years and older had not changed residence since birth. According to the Bureau of the Census' Current Population Survey, between 18.6 percent and 21.0 percent of the total population had moved to a different house within any given year since 1948; between 5.6 percent and 7.1 percent migrate to a different county annually, and about half that number to a different state. As far as can be inferred from the rather unsatisfactory figures, there has been little rise or fall in internal migration rates in the United States over the past 100 years, but a sharp increase in average distance of move. In any case, the persistent pattern of high mobility reinforces the belief that the footloose behavior of Americans is not a transient frontier phenomenon.

In recent years the outlines of a rather complex lifetime cycle of migration for the typical American citizen appears to be emerging. After a childhood during which his family may or may not move, and upon graduation from high school, the American may change residence one or more times as a result of marriage, military service, or college education. Then, with or without such an intervening phase in early adulthood, the adult participates for forty years or so in the labor force, during which time change in employment may prompt change in residence. In addition, continuing affiliation with a large corporation may mean shifts from one of its plants or office locations to another. Marriage and divorce

---

29 For useful discussions of the historic and geographic migrational patterns within the United States, consult Conrad Taeuber and Irene B. Taeuber, *The Changing Population of the United States* (New York: John Wiley & Sons, Inc., 1958), pp. 92–111, and Donald J. Bogue, *Principles of Demography* (New York: John Wiley & Sons, Inc., 1969), pp. 776–800.

30 One of the first serious attempts to confront this question is Larry H. Long, "On Measuring Geographic Mobility," *Journal of the American Statistical Association,* 65, No. 331 (1970), 1195–1203. This article tends to support the assertion that Americans are more mobile than citizens of other lands.

often bring about migration, especially for women. After retirement, the individual may elect to move his household to an area he considers attractive or comfortable. Each of these forms of migration produces important geographic effects.

The more casual and ephemeral movements in which Americans engage are poorly documented; but their aggregate volume, if it could be quantified as mileage or hours in transit, is enormous and probably well exceeds the level that could be recorded for any other advanced, nominally sedentary country.[31] Included here are several kinds of patterned, cyclical trips and a miscellany of one-time or irregular travel experiences. Among the former, the daily journey to work bulks largest. Only a tiny minority of American workers live within walking range of place of employment; and among the commuters, the 1963 Census of Transportation reports, 24 percent travel 11 miles or more to their destination and an appreciable number 30, 40, or more miles, generally by private auto. In addition, there are millions of weekend trips to vacation cottages or resorts or trips by college students from home to campus, innumerable delivery routes in retail and wholesale enterprises, and the complex odysseys of salesmen and business executives. Less repetitive in pattern are long-distance journeys by tourists and members of the military at home and abroad and a great volume of local (up to 50-mile) excursions to see friends, relatives and seek cultural or athletic diversions. In fact, the state of physical rest seems to be the exception.

Even though statistical details on various forms of American mobility can be elusive, the theme of dynamism is emphatically inscribed upon our settlement morphology. The highway has become *the* controlling element, the most vital social organ in contemporary life. No longer is it simply a bumpy, dusty nuisance linking densely settled places; it has been transformed into a spinal column, with cities and almost all else more and more dependent upon it for sustenance and growth.[32] The manner in which mountains and rivers are moved and great swaths slashed through cities with little heed to financial, social, or historic cost to make room for new highways testifies to the priority of values. (A century ago the same story could have been told about the imperious railroad.) The basic drives and motifs of a culture can be most clearly read along its growing spatial edges, where it is essentially uninhibited by the physical and social detritus of the past. This is perhaps the fundamental appeal of the settlement frontier of pioneer days to the social historian and cultural geographer. Today it is the suburb and ribbon settlements along our highways that speak most eloquently about

---

[31] On the significance of territorial mobility in general and the American situation in particular, see Wilbur Zelinsky, "The Hypothesis of the Mobility Transition," *Geographical Review*, 61, No. 2 (1971), 219–49.

[32] Some of the most thoughtful observations on the modern American highway and its attendant features are scattered through the issues of *Landscape*, in the form of brief essays by its former editor, John B. Jackson. See, for example, "The Abstract World of the Hot-Rodder," *Landscape*, 7, No. 2 (Winter 1957–1958), 22–27.

our essential selves; and even stronger clues are visible in the largely twentieth century cities that have sprouted in the relative emptiness of Texas, California, and the Southwest with their extraordinary dispersion and diffuseness and nearly total reliance upon the internal combustion engine—and the near extinction of the pedestrian.[33] There we find the logical physical embodiment of a society in which almost everyone, even the pauper, owns an automobile and the most significant rite of passage is the acquisition of a car or motorcycle after a childhood of roller skates, scooters, and bicycles. It is also a society where many aspire to motor-boats, skis, snowmobiles, and the private airplane: in short, speed is supreme.

Perhaps the most interesting physical expression of our theme of mobility is found in the large new suburban shopping center, with its sprawling parking lot or the developments that have mushroomed around important highway interchanges and airports, veritable autonomous cities, reminiscent of the almost totally self-contained large hotels of major metropolises—which, in turn, were spawned by railroad terminals. But there is also abundant evidence in the details of roadside enterprise within and between cities. The fact that the wheel has become an almost organic extension of the American's anatomy is illustrated by those indigenous inventions, the motel and the drive-in restaurant, cinema, and bank. Recently, drive-in post offices have begun to appear; and there are well-founded reports of drive-in churches and funeral parlors. In addition to the more conventional shops and services that stretch along the road for mile after mile, each with its parking lot, there are the many enterprises that would not exist without the auto, truck, or bus: filling stations, garages, accessory shops, new- and used-car dealers, drag strips, car laundries, and auto junkyards. Quite literally, travel has become an end in itself: the American never arrives; he is always on his way.

We have looked at only a small fraction of the evidence that could be amassed to prove that most Americans ardently subscribe to a value system granting lofty status to mobility in all its forms—travel, speed, haste, change—most obviously in physical space but in other dimensions as well. This is a proposition that will recur again and again in our further examination of spatial process and structure in the American cultural system.

---

[33] To a marked degree, the American city has been the creature of transport technology from the very beginning. See, for example, John R. Borchert, "American Metropolitan Evolution," *Geographical Review*, 57, No. 3 (1967), 301–32. One can define a sequence of urban morphologies identifiable with each major mode: first the city shaped by the needs of ocean or river transport, pedestrians, and horse-drawn vehicles; then the patterns engendered by railroads and, somewhat later, by streetcars; and currently the morphological response to auto, truck, and airplane. The basic point remains the same, however: the overwhelming importance in urban design of moving people and things. Cities tend to be marts and transfer points; all else is subsidiary.

The Theme of a Mechanistic World Vision

Another basic postulate upon which American culture is constructed is that the universe is a fundamentally simple, mechanical system subject to human control: this is the vision of the world as machine, one that can be improved and accelerated.[34] (Note the peculiar prominence of the term "fix" in the American vocabulary.) This belief applies not only to the cosmos in general, but to the various subsystems and territorial units within it, as well as to individual human beings. Although this theme is ultimately as meaningful for the human geography of the United States as the notions of individualism or dynamism, its spatial expression is rather less direct, so that it can be expounded here briefly.

A mechanistic model of reality leads quite readily to a series of related ideas: the desirability of keeping the machine and its components nicely adjusted and working at optimum efficiency; a cleanliness fixation; a highly elaborated pragmatism; hence a utilitarian attitude, with a corresponding denigration of abstract thinking and aesthetic considerations; a passion for the precise measurement of things; a love of bigness and growth; a strongly extroverted personality pattern; and a firm belief that brute nature can be subjugated and made to behave rationally for the benefit of its human masters. Only some of these beliefs call for further comment.

Within the adjustable world-machine envisioned by Americans, the human actor is a detachable, freely mobile, perfectible cog; and we achieve the convergence of our previous two themes with that under discussion. Eli Whitney's invention of the system of interchangeable parts, perhaps the first truly major American contribution to world technology, provides a neat paradigm for the larger pattern of reality as sensed by his compatriots: the interchangeable Yankee, highly versatile and movable, and eager to insert himself into the locus of maximum advantage to himself and thus, of course, to the system.[35] The phenomenal social, economic, and spatial mobility of the individual thus becomes a rational way for lubricating and putting into a state of frantic productivity all those gears, wheels, levers, springs, and pulleys that comprise the physical and social world. The pressures within the cultural milieu tend to mold people into flexible, adjustable, cheerful, conformable units for operation in the social as well as the economic sphere. If a machine is to work well, its parts must be washed, dusted, and carefully cleaned and polished; and for this reason, among others, we find an obsessional interest in personal cleanliness, which carries over

[34] This notion is developed with great power and depth in two volumes by Lewis Mumford, *The Myth of the Machine: I. Technics and Human Development* (New York: Harcourt Brace Jovanovich, Inc., 1967) and *The Myth of the Machine: II. The Pentagon of Power* (New York: Harcourt Brace Jovanovich, Inc., 1970).

[35] Daniel J. Boorstin, *The Americans: The National Experience* (New York: Random House, Inc., 1965), esp. pp. 3–48.

into the internal design of the American house. But it must also be noted that while the American is so concerned with the neatness of his person, clothes, and home, this interest fades past personal property lines, so that he feels almost complete indifference to the appearance of public spaces.

A mechanistic world is a quantifiable one, so that it is hardly surprising to find everything numbered and counted, or to note the ardent accumulation of statistics by a people who dote on digits. The observable universe is converted into dollars, hours, miles, acres, tons, decibels, input-output tables, foot-candles, wavelengths, millibars, TV ratings, grade-point averages, and a long list of other indices. It scarcely needs saying that the dominant themes and methods in American social science are set by the general cultural conditioning of its practitioners. In territorial terms, the rectangular land survey, already mentioned for its effectiveness in dealing with the craving for personal isolation, has the additional virtue of numerically coding even the smallest tract with its unique combination of letters and numbers, so that both size and location can be identified almost instantly. The great majority of American cities are wholly or partially laid out on a rectangular grid; and many of them have their major thoroughfares named in numerical or alphabetical order. The practice of methodically numbering all buildings within a city (and all rooms within buildings) by distance and direction from a baseline or point of origin seems to be an American innovation, probably one initiated in Philadelphia. It is now almost universal within the United States, but has been adopted and executed with equal rigor in relatively few places outside North America.

A reverence for number and quantity combines with the pervasively dynamic quality of the American cultural system to produce a fervent worship of both growth and bigness. The theme of giantism—paralleled in modern times only by the Russian passion—is constantly obtruding in American behavioral patterns and the visible landscape. In many American cities, the skyscraper makes more sense as symbol than as rational investment.[36] Humorous braggadocio and the tall story seem indigenous, and Texan expansiveness in myth and jest is a quintessential expression of this impulse. There is an irresistible urge to achieve—and proclaim—the quantifiably superlative—the biggest, highest, costliest, loudest, or fastest—frequently without any dollar-and-cents justification. This compulsion cannot be ignored in analyzing the economic or population geography of the nation; indeed it merits book-length treatment in its own right.

The American value system casts its members in the roles of Promethean heroes bringing light and imposing order upon a somewhat unruly, but ultimately domesticable, wilderness. As subjugators and transformers of nature, Americans tinker with or rebuild its machinery and make it run faster and more smoothly. In the short run, this strategy

---

36 Jean Gottmann, "Why the Skyscraper?" *Geographical Review*, 56, No. 2 (1966), 190–212, and Larry Ford, *The Skyscraper: City Structure and Urban Symbolism*, doctoral dissertation, University of Oregon, 1970.

has generally succeeded, at least in bookkeeping terms, even though the long-term resolution of the ecological problems of man in North America is cloudy at best. The march of settlers westward across the continent can be viewed as a series of environmental traumas or conflicts, in each of which the modern American has won the immediate decision through a technical knockout. The principle in every case has been that the land must be made economically profitable, to respond quickly and effectively to human endeavor; it is a simple commodity to be sliced, cubed, sold, processed, and squeezed of its intrinsic wealth.

Through a fluke of environmental fate, the culture hearth within which these ideas were first fully hatched and tested was the northeastern United States, an area reasonably analogous in physical characteristics with the Northwest European source area. Few concessions were necessary. Within the humid subtropics of the American South, on the other hand, the traditional modes of rural economy did not work very well. But instead of coming to ecological terms with an unfamiliar land, the Southerner persisted in his Promethean ways and inflicted upon the region a totally inappropriate mode of cultivation. In the case of cotton and tobacco, the early profits were reassuring, even though growing conditions, especially for the former, were marginal; but the ultimate cost in environmental and social havoc was incalculably great. It is only in the twentieth century that ways have been found to make Southern land productive for long periods at fairly high levels; but, again, through a synthetic technology willfully imposed by men who are still not psychologically at home in the region.

In less dramatic form, the minor crisis of the encounter with thick-sodded humid grasslands in the Middle West was overcome with the aid of technology; and the land was "broken in" promptly and efficiently. The chronic environmental tensions in the Great Plains and further west within more arid regions have never been fully resolved. Instead there has been a series of technological adventures, some of them failures, others temporarily successful, whereby the American conquistador has tried to tame the land. On an even larger scale, there is growing concern over the long-term biological and psychological results for the American population as a whole, and not just the physical impact upon land, water, and air, of the runaway explosion of a technology that ignores the habitat, or else treats it in a highly simplistic fashion.

The Theme of Messianic Perfectionism

Perhaps the most extraordinary facet of the American national character is the power and durability of an idea only weakly, fitfully developed in other cultures: the notion that the United States is not just another nation, but one with a special mission—to realize the dream of human self-perfection and, in messianic fashion, to share its gospel and achievement with the remainder of the world. This moral expansiveness (some would call it "moral imperialism") exists over and beyond the usual flexing of economic or military muscle that all modern nation-states of any size seem to indulge in quite automatically in response to some general law. If it transcends standard nationalism, neither is it just the normal afterglow of revolutionary fervor. Indeed the American Revolution was much tamer in its initial ideological content than other modern revolutions have been. But if there is a predictable time-decay function for the Islamic, French, Russian, and presumably the Mexican, Gandhian, Maoist, Castroite, and lesser political and religious convulsions of recent centuries, as the curve representing zeal for converting the heathen droops steadily lower each year, quite the opposite seems to have happened with the American sense of mission.

From the very beginning of serious settlement along the Atlantic Seaboard, potent strains of a visionary perfectionism were mingled with more mundane economic and political considerations, and not only in the New England colonies. In this errand into the wilderness, there was a quite conscious attempt to seek out or build a New Zion—later the New Athens or New Rome—the City upon the Hill, or quite literally, an earthly Eden with a Manifest Destiny.[37] It is not too extravagant to claim that collectively, the country thinks of the land as a literal, real-world utopia. Thus it is that many a citizen can quite solemnly speak of "God's Country" or the "American Way of Life." Even the term "Americanism" carries with it a sense of uplift, special earnestness, and quasi-religious connotations; no other language has a parallel term.

Insofar as cultural geography is concerned, this is a major, but generalized, theme more important as basic premise or mood than as specific, mappable expression. But the American penchant for messianic perfectionism does attain tangible form in at least two ways, neither of which has received the geographic analysis it deserves. If one might debate the propriety of treating the entire nation as a utopian experiment, the case is clear for early New England and other colonial settlements, notably Penn's Holy Experiment. Some scores of churchly and secular utopian colonies have been founded in the United States from the

---

[37] This idea is implicit in many writings on the character of life and thought in the United States but is seldom advanced explicitly. Among the more noteworthy statements are Hugo Münsterberg, "The Spirit of Self-Perfection," *The Americans* (New York: McClure, Phillips & Co., 1905), pp. 347–64; Jean Gottmann, "Prometheus Unbound," *Megalopolis: the Urbanized Northeastern Seaboard of the United States* (New York: Twentieth Century Fund, 1961), pp. 23–79; Charles L. Sanford, *The Quest for Paradise: Europe and the American Moral Imagination* (Urbana: University of Illinois Press, 1961); and J. Wreford Watson, "Image Geography: the Myth of America in the American Scene," *Advancement of Science*, 27 (1970–1971), 1–9. The argument is especially fully and eloquently developed by Sanford, and one of his central statements is well worth quoting:

> Liberals and conservatives alike have been moralists, forever pitting themselves...against the fancied evils of life, forever demanding changes, improvements, cure-alls. Since Colonial days, in fact, there has been a marked tendency to regard America as having greater moral obligations than other nations and therefore to consider her sins more heinous than those of other nations.
> An unfortunate result has been much exaggeration of America's deficiencies by patriotic citizens and an equally unfortunate exaggeration of her innocence and uniqueness in terms of moral (material, political, economic) superiority. By their great prestige as culture heroes, Franklin and Jefferson contributed significantly to a national superiority complex which alternated between an isolationism which would retreat behind a *cordon sanitaire* of protecting oceans to a hermetically sealed paradise in the Mississippi Valley, and a messianic internationalism which would make the world over in the American image. (p. 134)

eighteenth century onward.[38] The movement may have peaked during the last century, but new, ever more inventive utopias are still being formed. Usually quite short-lived and practicing some form of primitive communism, such visionary schemes were not unknown in Europe and in other neo-European areas; but their incidence and their impact upon national thought has been far greater here than elsewhere. In one instance, the Great Basin Kingdom of the Mormons, the enterprise has flourished, endured, and had a decisive effect upon the human geography of a truly extensive tract of land.[39] Although these utopian adventures account, in sum, for only a small fraction of the population, they are deeply symptomatic—a kind of psychodrama, acting out the daydream haunting the American cultural soul.

In their choice of place names, Americans have also revealed various facets of their world outlook, including a vigorous strain of self-improvement and upward aspiration. Almost any corner of the country will do, but a casual scanning of the Pennsylvania map yields the following more or less inspirational town names:

Acme	Frugality	Philadelphia
Arcadia	Harmony	(City of Brotherly Love)
Concord	Independence	Pleasant Unity
East Freedom	Industry	Progress
Edenville	Liberty	Prosperity
Effort	New Hope	Salem (Peace)
Enterprise	New Jerusalem	Unityville
Freedom	Paradise	Zion

But it is in larger, less concrete manifestations that the force of this faith in America's unique burden has been greatest. The pervasive feeling of optimism and enthusiasm, one that does appear in the normal behavior of most individuals, reflects this inner faith; and it has supplied much of the drive toward extraordinary exertions in the economic arena or in the physical manipulation of the habitat. The remarkable, unabating vigor of American religious missionary activity abroad, usually dispensing Americanism, covertly or otherwise, along with a particular brand of Christianity, bubbles up from the same cultural wellspring.[40] American philanthropies—educational, technical, medical, and charitable, including the Peace Corps—exert strenuous efforts in almost every corner

---

[38] Among the more useful accounts are Charles Nordhoff, *The Communistic Societies of the United States from Personal Visit and Observation* (New York: Harper & Row, Publishers, 1875) and Mark Holloway, *Heavens on Earth; Utopian Communities in America, 1680–1880* (New York: Dover Publications, Inc., 1966).

[39] Donald W. Meinig, "The Mormon Culture Region: Strategies and Patterns in the Geography of the American West: 1847–1964," *Annals of the Association of American Geographers,* 55, No. 2 (1965), 191–220, and Richard V. Francaviglia, "The Mormon Landscape: Definition of an Image in the American West," *Proceedings of the Association of American Geographers,* 2 (1970), 59–61.

[40] "A geography of missionary operations, both Catholic and Protestant, is needed for a proper evaluation of Christianity's role as a carrier of Western culture to the non-Western world, especially since 1800. Unfortunately, a comprehensive

of the globe. In a quite fundamental, but often unrealistic way, the basic postulates and decisions in American foreign policy have been affected by the enduring conviction of an ordained vocation.

On the domestic scene, the intensity of various ameliorative movements—for temperance or prohibition, for the abolition of slavery, for civil rights, for women's suffrage—and of the counter-movements has reached a pitch beyond that justified by the immediate practicalities and may attest to a streak of fananticism. In the geography of each of these, their regional evolution through space and time, and the interaction with other social and cultural items, much could be learned about the larger issues of American cultural geography, but little serious work has been attempted to date.

Finally, the American quest for a secular paradise has had specific implications for the physical scene of much interest to the geographer. Recent studies have richly supported the suggestion of a pastoral vision as a major component of the American Dream.[41] The United States has been regarded, subconsciously by most, but with a dawning awareness by writers and artists, as a bucolic Middle Landscape, a humanized, parklike blending of the best of wild nature with civilization's blessings, neither wilderness nor city, but a blissful balance between the primitive and the urbane. The self-sufficient single-family farm tilled by virtuous Jeffersonian yeomen has been the physical embodiment of this ideal. And a persistent reverence for this rural Eden and the only slightly less idyllic small town still lingers in the folk memory. The strangely ambivalent attitude toward wilderness derives from this vision, as does the chronic allergy toward the completely urban life. One might also conjecture that the almost intuitive westering urge of the American migrant, a momentum that has not spent itself more than seven decades after the official closing of the frontier, or the peculiar obsession with East Asian affairs cannot be fully explained as the search for economic advantage or for physical delights, neither of which is necessarily fully realized by these pursuits. We may be witnessing the pursuit of an ever-receding mythic landscape, a pastoral perfection that has always lain to the westward, first for the European, and later for the American.

---

mapping of the extent and intensity of Christian missionary activity does not exist." David E. Sopher, *Geography of Religions* (Englewood Cliffs, N.J.: Prentice-Hall, Inc., 1967), p. 70. And there is indeed a striking paucity of serious geographic work on this promising geographic topic. Most of the important cartographic and other endeavors in this sector are noted in Hildegard Binder Johnson, "The Location of Christian Missions in Africa," *Geographical Review*, 57, No. 2 (1967), 168–202. This essay is one of the most substantial contributions to date on missionary geography.

41 Among the better presentations of this theme are Henry Nash Smith, *Virgin Land: the American West as Symbol and Myth* (Cambridge, Mass.: Harvard University Press, 1950); Sanford, *The Quest for Paradise;* Marx, *The Machine in the Garden;* Peter J. Schmitt, *Back to Nature: the Arcadian Myth in Urban America* (New York: Oxford University Press, 1969); and Watson, "Image Geography."

PART 2 *spatial expression*

In the two chapters that comprise Part Two, our approach becomes more conventionally geographic as we explore some explicitly spatial attributes of American culture. A general statement concerning the basic spatial processes operating within the cultural system of this country will be attempted in chapter 3. First, we must begin by backing off and examining some fundamental issues in the study of culture and of cultural geography in particular. Then, after tying these general items into the specific conditions of the American scene, we will treat briefly two sharply contrasting sets of phenomena—the American house and the geography of American religion. The purpose of this exercise is to show how basic cultural processes achieve specific geographical expression, and also to document the relevance of the critical traits of national character discussed in the preceding chapter. The next chapter takes up one of the more important and interesting, mappable manifestations of the processes in question—the culture area—and examines the identity, origin, structure, and significance of American cultural regions and subregions, past and present.

There is some precedent for the initial chapters of this volume, although it must be sought largely in the nongeographical literature. There appears to be none for chapters 3 and 4, which represent, however haltingly, an effort to treat the spatial behavior of American culture systematically and with some rigor. Therefore some general definitions of concepts are in order, as is a confession of the inadequacy of our knowledge and methods.

CHAPTER THREE

Process

Explaining the Spatial Aspects of Cultural Change

What processes have been most influential, within the total cultural system, in shaping the geography of the country? And how have they operated? The question may also be rephrased to read: How and why has American culture changed through time and space? The problem has already been approached in chapter 1 in which we described, or speculated about, the ways in which a distinctive American culture came into being during the colonial and early republican periods. A major part of the answer was seen to be the spatial transfer and subsequent interaction of selected people and traits from foreign sources. The simplest way to attack the questions just posed might appear to be to pursue this strategy further—to discuss the spatial transfer of objects (people, things, and ideas) within the United States and a continuing interchange with other parts of the world.

In actuality, however, several distinct processes have been at work. These are:

1. The selective transfer of immigrants and cultural traits from the Old World;
2. The interaction of the newcomers among themselves and with new habitats in several early cultural hearths;
3. Differential participation by various groups in the advance of the settlement frontier from these cultural hearths;
4. Differential mobility of different groups of people during the post-pioneering period;
5. The spatial diffusion of a great range of specific innovations;
6. Deep structural change in society and culture that is expressed at different times and rates in different tracts.

All these mechanisms of change have been, or will be, noted (though in rather different order); but special attention should be accorded the one that may be the most important, yet the least amenable to direct observation: the deep structural changes experienced by a society and culture as a community evolves upward from a relatively primitive set of conditions toward an ever more complex civilized existence. The vast question of the degree to which the evolutionary paths of developing societies are followed in blind obedience to fundamental historical laws or, on the contrary, are coincidental in nature or the result of contacts among different societies, is one of the most difficult and controversial facing the cultural anthropologist.[1] Thus a final decision may never be reached as to whether the suggestive similarities between the social and ecological structures of early Middle Eastern and Middle American civilizations constitute two special cases of the same universal and inevitable progression or are simply coincidental.

In any case, if one were to amass all possible data on local inventions, the diffusion of innovations, interaction with the local habitat, the spatial movement of people and influences, and all the other discrete events that contribute to culture formation, there would still be a large, unexplained residuum. Much of what has happened in the slow character-building process in the culture history of a locality would seem to be transacted at the unlit subterranean levels of consciousness, as a series of extremely gradual, subtle shifts in modes of thinking, feeling, and impulse in response to basic alterations in socioeconomic structure and ecological patterns. There is no reason to believe that such has not also been the case with the United States and its various subregions, even within the relatively brief time this society has existed.

Some Basic Cultural Propositions

Before we can begin exploring the how and why of spatial and temporal shifts in American culture we must take a hard definitional look at what is being studied: the concept of culture. Only within the past few decades have students of mankind begun to recognize the existence of an entity called "culture," something within, yet beyond the minds of individual human beings.[2] It is a very large, complex assemblage of items, which, taken together, may be as important a variable as any in explaining the behavior of individuals or societies or the mappable patterns of activities and man-made objects upon the face of the earth. The late emergence of any semblance of a "science of culture" can be

---

[1] Carroll Quigley, *The Evolution of Civilizations: An Introduction to Historical Analysis* (New York: The MacMillan Company, 1961); and William T. Sanders and Barbara J. Price, *Mesoamerica: The Evolution of a Civilization* (New York: Random House, Inc., 1968).

[2] For extended discussions of the concept of culture and its significance, see Leslie A. White, *The Science of Culture* (New York: Farrar, Straus, & Giroux, Inc., 1949); and Alfred L. Kroeber, *The Nature of Culture* (Chicago: University of Chicago Press, 1952).

attributed in part to the extraordinary difficulty of observing and objectively measuring the characteristics of so complex and elusive a phenomenon. The idea that the traditional ways of thinking and acting of one's group are not absolute and that there is some coherence in the seemingly chaotic kaleidoscope of beliefs and customs of alien societies required a bold leap of the anthropological imagination.

The history of scientific thought also helps account for our inability to offer a fully rigorous definition of culture or to suggest many firm ideas about the structure of cultural systems or the laws governing their behavior through space and time.[3] The physical and chemical properties of inorganic matter are the most obvious, measurable items for the curious mind searching after some underlying order in the universe; and their study initiated formal modern science, as we know it. The methodical observation of plants and animals appeared soon after as another scientifically respectable endeavor, despite the greater difficulties involved. Very much later, the human mind began to be inquisitive about itself and the properties of our nervous system and personality, and the science of psychology was born. It is when the scientist approached groups of things in complex interaction that both observation and analysis posed the most formidable challenges. Sociology and political science are new, and their achievements relatively modest. So are the "scientific" approach to history[4] and the study of ecology, that is, the "societal" aspects of plants and animals coexisting within specific habitats.[5] The systematic understanding of the culture of human groups calls for sophisticated techniques of an even higher order. Not the least of the problems is achieving adequate objectivity, a relatively minor matter in the physical and biological disciplines. The student of culture must somehow strip himself of his native preconceptions. It is difficult enough to do so when looking at alien folk; it entails a near miracle when investigating one's own culture, as in the present work.

Much ink has been spilled in the effort to reach a satisfactory definition of culture. Perhaps the most successful to date is that offered by Kroeber and Kluckhohn in 1952 after an exhaustive critique of the literature:

Culture consists of patterns, explicit and implicit, of and for behavior acquired and transmitted by symbols, constituting the distinctive achievement of human groups, including their embodiment in artifacts; the essential core of culture consists of traditional (i.e., historically derived and selected) ideas and especially their attached values; culture systems may, on the one hand, be con-

3 Of the many recent works on the history of science, the most useful would appear to be J. D. Bernal, *Science in History,* 4 vols. (Cambridge, Mass.: M.I.T. Press, 1971).

4 David S. Landes and Charles Tilly, eds., *History as Social Science* (Englewood Cliffs, N.J.: Prentice-Hall, Inc., 1971).

5 The emergence of the fields of biological and social ecology is thoughtfully reviewed in Otis Dudley Duncan, "Social Organization and the Ecosystem," in R. E. L. Faris, ed., *Handbook of Modern Sociology* (Chicago: Rand McNally & Company, 1964), pp. 37–82.

sidered as products of action, on the other as conditioning elements of further action.[6]

The full exegesis of this statement could, and did, require a full volume; and some cultural anthropologists would take exception to all or part of the definition. But all would agree that culture is an assemblage of learned behavior of a complexity and durability well beyond the capacities of nonhuman animals. Following the Kroeber-Kluckhohn formulation, culture can be regarded as the structured, traditional set of patterns for behavior, a code or template for ideas and acts. It is highly specific to each cultural and subcultural group, and survives by transfer *not* through biological means but rather through symbolic means, substantially but not wholly through language. In its ultimate, most essential sense, culture is an image of the world, of oneself and one's community.

It will be helpful to spell out several general attributes of culture implied in the definition above that are of importance to geographers and other students. First of all, culture is indeed an exclusively human achievement. In fact, it is *the* critical human attribute, the one exclusive possession that sets mankind far apart from all other organisms. With the appearance of cultural behavior at least one million years ago during the organic evolution of man-like creatures, true human beings can be identified. The capacity for abstract thinking, for the mental juggling of symbols and inventing of ideas, is something no other species has attained, as far as we can tell.[7] Given this ability, enormous stores of information and ideas can be accumulated, sorted out, retained and bequeathed to later generations, and added to; an unlimited variety of recipes for thinking and behaving can be assembled from this vast larder of knowledge, then adopted, revised, or abandoned. Thus the pace of cultural evolution far outruns that of the slowly mutating germ plasm. Such plasticity and inventiveness, certainly not any inherent physical prowess, accounts for *Homo sapiens* having become the dominant species on this planet. Implicit in all this is a very great variability in modes of behavior as between groups, places, or periods, much more than might normally be expected of members of the same species. At the same time, there is also a strong intrinsic conservatism in many phases of a culture, so that only a small fraction of all combinations have ever been realized.

The power wielded over the minds of its participants by a cultural system is difficult to exaggerate. No denial of free will is implied, nor is the scope for individual achievement or resourcefulness belittled. It is simply that we are all players in a great profusion of games and that in each cultural arena the entire team, knowingly or not, follows the local set of rules, at most bending them only slightly. Only a half-wit

---

6 A. L. Kroeber and Clyde Kluckhohn, "Culture, a Critical Review of Concepts and Definitions," *Papers of the Peabody Museum of American Archaeology and Ethnology*, 47, No. 1 (1952), 181.

7 Try teaching even the brightest chimpanzee or dolphin the meanings of the words "mother," "yesterday," "kilometer," or "flag."

or a fool would openly flout them. But as in chess, the possibilities for creativity and modulation are virtually infinite. It is enough to have experienced "cultural shock," the sudden immersion in another culture without special briefing, or the almost equally painful reentry into one's own community after such an episode, to realize that many of the habits one regards as natural or logical are so only for one's own group. Most of the norms, limits, or possibilities of human action thus are set as much or more by the configuration of the culture as by biological endowment or the nature of the physical habitat. It is futile to evaluate or give relative weights to the various determinants that shape the affairs of men, since none can operate without the others. The survival of a cultural pattern is dependent upon inherited human capabilities and upon a certain range of habitats, and its functioning is subject to whatever constraints are intrinsic to these; but, on the other hand, cultural expression also makes a deep impact upon the outer world. Thus culture is a powerful agent, but always in complex interaction with other powerful systems.

A cultural system is not simply a miscellaneous stockpile of traits. Quite to the contrary, its many components are ordered. Moreover, the totality of culture is much greater than the simple sum of the parts, so much so that it appears to be a superorganic entity living and changing according to a still obscure set of internal laws. Although individual minds are needed to sustain it, by some remarkable process culture also lives on its own, quite apart from the single person or his volition, as a sort of "macro-idea," a shared abstraction with a special mode of existence and set of rules. This point becomes clearer if one examines some specific cultural complexes that share this attribute of superorganic existence with the total culture. Thus an economic system—*vide* the famous "Invisible Hand" of Adam Smith—has been perceived to evolve and act according to its own private code, at least in pre-Keynesian times, without the effective intervention of its participants. Languages constantly change, though at variable speeds, in accordance with complex rules we are only slowly beginning to grasp, but utterly without calculation or effort by their speakers. Any sensitive linguistic observer who has watched the dizzy pace at which American English has altered during the past generation can testify to a sense of lying helpless in the path of large anonymous forces. Similarly, there is a distinctive personality to be recognized in almost any viable organization—church, college, army, corporation, or government bureau—that almost literally lives and breathes, persists and develops, quite independently of the personal sentiments of its members.

The nation-state idea is perhaps the neatest illustration of the transpersonal character of cultural systems; and the origin, growth, and perpetuation of the idea of a United States of America is a superb example. Whatever the genesis of the idea, it was certainly not the conscious fabrication of any identifiable cabal of nation-builders, but rather a spontaneous surge of feeling that quickly acquired a force and momentum of its own. This idea was so powerful that millions of men were ready to sacrifice their lives for it. Individuals who entertain the nation-state idea are born and die, and some may even have doubts or reservations; but the idea marches on, quite clearly beyond the control of anyone. Even if it were a matter of dire necessity, it seems impossible to devise any program, excluding mass annihilation, whereby the idea of a United States or a Russia, France, or Germany could be disinvented.

The structure of cultural systems is rather loose and open. If we regard culture as a system in the technical sense of the term, it is rather special by reason of both complexity and sheer size. Probably the only other system of greater magnitude is that including all interacting sub-systems on and near the face of the earth that comprise total terrestrial reality and the subject matter of geography. There may be certain quintessential ideas and practices that cannot be tampered with without profoundly revising the larger cultural pattern; but, in the main, the total structure can absorb much change or contamination, including addition or subtraction of elements in specific departments of culture without great impact upon the whole. Thus two centuries of radical technological and economic change have not basically revised the structure of the American family—at least not yet. And the near-disappearance of men's straw hats, electric trolleys, or Spencerian handwriting seems to have had minimal effect upon basic American life patterns. But if one could imagine anything as unimaginable as a mass conversion to ascetic Buddhism, the substitution of the Arabic language for English, the abolition of the achievement motive, or the adoption of a joint family system, the reverberations all through the cultural matrix would be fast and shattering.

Even among so-called "primitive folk," a single cultural system encompasses an enormous range of information. Each cultural group has a certain common fund of traits—a full count of the individual bits of information would probably run well into the millions—that is acquired, usually quite unconsciously, during the early months and years of childhood. But, in addition, there are any number of special groups or activities, ranging from a half-dozen or so in the simplest of societies to literally hundreds of thousands in the most complex, each with its own distinctive subculture. Even in a Micronesian village or a band of Bushmen, no single person commands the full complement of traditional practice and lore. In the United States, even the most diligent and brilliant of scholars could not begin to master more than a minute fraction of the aggregate cultural package.

The problem of how to take an inventory of all traits or complexes that make up the total system, or how to classify the full range of subcultures and other major dimensions present within a given community, has not been solved. But the situation can be indicated roughly in diagrammatic form (Figure 3.1). Consider the full rectangular solid as representing the total culture, with one dimension equivalent to the range of traits and complexes that make up the totality of any culture or subculture. A second dimension (here the vertical one) represents an additive (and overlapping) set of subcultures: for example, males and females, farmers, ditchdiggers, Presbyterians, mountain climbers, convicts, Freemasons, bowlers, and drug addicts. (The number of strata

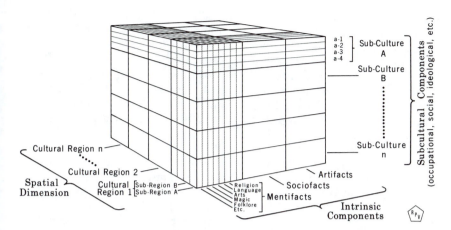

FIG. 3.1.   *A schematic, three-dimensional representation of cultural systems.*

shown in the diagram is, of course, highly schematic.) The third dimension indicates variability through space, that is, the set of cultural regions and subregions that one can identify within the territorial range of the cultural system in question. Further dimensions—especially time—might be added to this scheme, but not without transcending the possibilities of graphic representation. Note that if the cube is sliced either horizontally or vertically (at an angle normal to the spatial face), the result is a regional or subcultural parcel of cultural phenomena that run the gamut of human ideas and practices: for example, courtship, facial expression, superstition, pronunciation pattern, motor skills, social etiquette, and burial customs.

For our purposes, it will suffice to adopt one of the simplest ways of categorizing the components of a culture, among the almost limitless array of possible schemes. This is a tripartite classification into artifacts, sociofacts, and mentifacts.[8] Artifacts are those elements of culture that are directly concerned with matters of livelihood or, somewhat more broadly, the entire technology of supplying wanted goods and services. The variety of artifacts can barely be suggested: all tools, weapons, and other man-made objects; manufacturing in all its many aspects; the shelter system; the production of food and drink; the transportation system; medicine; property-holding and land-use systems; clothing; and many other phenomena. Sociofacts are those phases of the culture most directly concerned with interpersonal relations: kinship and family systems; political behavior; education; social etiquette; voluntary organizations; reproductive behavior; child rearing; and a host of others. Mentifacts are basically cerebral, psychological, or attitudinal in character, and include religion, along with other ideological baggage, magic and superstition, language, music, dance, and other arts, funerary customs,

[8] Julian S. Huxley, "Evolution, Cultural and Biological," *Current Anthropology,* 7 (1966), 16–20.

folklore, the basic value system, and abstract concepts of all sorts. In a sense, the mentifactual is the innermost, least mutable, "holiest," and most precious segment of the culture—the glue holding together the entire cultural mass and setting its tone and direction.

In practice, it is hard to find any single facet of culture that is purely artifactual, sociofactual, or mentifactual. These arbitrary categories are interdependent to a marked degree. For example, house design and construction, which might appear to be wholly technological or artifactual, is, in fact, closely associated with the nature of the family and social system and so also with religious or cosmological ideas. In any case, if the total culture can be seen as a loose, yet somehow structured, assemblage of an almost innumerable set of elements, viewed from another angle, it is also a package with many subcultural compartments, each with a decided amount of autonomy, and to many of which the individual may belong simultaneously. To make this thought more concrete, consider the single man, the microcosmic building block of a larger cultural universe, who carries his own unique collection of cultural attributes, and may also be a participant in many subcultural groups. Imagine someone who is, among other things, a Czech-American Lutheran plumber, a member of the VFW, an ardent Cleveland Indian fan, a radio ham, a regular patron of a particular bar, and a member of a car pool, the local draft board, the Book-of-the-Month Club, and the Republican party, and a parent whose son attends a particular college. Each of these subcultures will tend to have its own array of gear and physical arrangements, spectrum of economic and social beliefs and practices, cluster of abstract concepts, and, not least important for our purpose, distributional spread in physical space. The one man and his friends and associates move through many worlds. It is only by taking into account the relevant multiplicity of components and dimensions at the particular scale chosen for observation, and also the fact that they are changing through time, that a realistic understanding of a given culture can be reached.

It has been implied that virtually the entire range of human activities might be regarded as the acting out of various cultural systems. Consequently, one might make the technical claim that cultural geography is equivalent to the entirety of human geography. But this sort of academic imperialism is impractical. For most purposes, it is more efficient and profitable, at least in the short run, to extract major subsystems from the total culture, then treat them independently of other aspects of the larger system, holding other things constant. This has been done with conspicuous success in the case of the economic, demographic, and political subsystems, especially when the analyst restricts himself to a single reasonably homogeneous cultural realm. It is even convenient, when working at the international scale, to assume universal economic, demographic, political, and perhaps other principles that override and suppress local cultural peculiarities. One result is that the normal domain of the cultural geographer includes only those items that are overtly ideological or traditional, such things, for example, as language, religion, ethnicity, the settlement landscape, or diet, or whatever is least susceptible to rigorous systematic analysis and the prompt distillation of general propositions.

As it happens, we confront here a profound, but as yet poorly understood, methodological dilemma. Each body of human geographic phenomena stands at the intersection of two large systems: that of its local culture and that system of ideas or relationships governing the branch of knowledge in question. Thus the geography of land use in the United States can be approached either as one of the outward ways in which the indigenous culture is manifested, or simply as the local expression of a group of universal economic principles that we have had some success in capturing. At the moment, the latter strategy is rather more immediately productive, although an ideal plan would find some way of merging both perspectives. On the other hand, if we seek to understand the physical morphology of rural settlement in this country, it is more expedient to work within the context of the national cultural pattern rather than to appeal to universal postulates, since we still have only rather fuzzy notions of what those might be.

As is true for any geographical phenomenon, the question of scale is critical in dealing with cultural systems. As used here, "scale" is not simply the problem of the fineness of the detail under scrutiny or, as with map scale, the ratio between the real object and the abstraction. There is the even thornier matter of bounding the cultural facts to be studied. How large a group or territory is to be observed? Cultural systems are loose, open affairs with hierarchical tendencies and with few sharp natural boundaries. (We ignore such trivial exceptions as the island or oasis community isolated over long periods.) There is an unbroken continuum of cultural units—of boxes within boxes, so to speak—extending downward from the whole of humanity, living, dead, or yet to be born, to the single contemporary individual, and each of them is unique. Obviously, neither extreme is too useful in an operational sense. (There is also the related question of temporal duration. When is a culture born, when does it die, and how do we segment its lifespan into meaningful periods?)

Fortunately, we do have some fairly distinct, significant levels within the nested hierarchy. In the contemporary world, the nation-state—especially, the robust, strongly integrated, long-established nation-state—is a prominent, if not airtight, cultural compartment. The tactic of analyzing the cultural geography of the United States, which this volume attempts to do, is quite defensible; but the porosity and multidimensionality of culture is such that we must not fall into the trap of thinking that the official boundaries of the United States rigorously define the American culture area. There are ambiguous zones both within and outside the international frontiers. It is extremely difficult to say precisely who is or is not an American in many cases (the expatriate? the recent immigrant? the aborigine?); and many Americans are members of subcultures with foreign linkages. Rotarians, spelunkers, stamp collectors, gourmets, Baptists, movie buffs, student radicals, and xylophonists all have soulmates abroad.

There are other cultural aggregations cohesive enough to commend themselves to our attention. On the broadest territorial scale, we find the large cultural realm, for example, the Islamic realm or the European.

At the subnational level, one may look at ethnic groups and certain distinct subnational "culture areas," which have many of the traits of ethnic groups but do not quite aspire to nationhood, the critical feature of the full-fledged ethnic group. The student can even descend to the small tribe or the single village or band. In this work, we are concerned only with the culture of the American nation-state and of some of its more distinct culture areas, units ranging in size from a few thousand to some hundreds of thousands of square kilometers.

The one attribute of cultural systems that most particularly interests us is the fact that they almost always have spatial dimensions, that is, they exist within certain localities. Moreover, the general nature of the locality seems somehow to shape the culture and, in turn, to be influenced by it. In the strictest sense, the spatial component of culture is incidental. As a cerebral entity, a culture may flourish, move about, and propagate itself solely within the heads of a number of footloose individuals. Such extreme cases do occur, of course, but normally the facts of location and the processes of interaction with other localized or spatially structured phenomena do matter greatly. In fact, the territorial dimension is strong enough that it seems fitting to accord the regional aspect of culture an importance rivaling the technological, social, or ideological. This statement is especially valid when representatives of a culture have become deeply rooted in a specific place.[9]

A particular culture, or combination of subcultures, helps impart to an area much of its special character and behavioral design—which, fundamentally, is what geography is all about. Conversely, the character of a given place may be a strong formative influence in the genesis of a particular culture, which is the ultimate concern of the cultural anthropologist. This is especially so, it must be stressed again, after a cultural group has become established in a certain tract, or when such a group is transferred en masse to another tract. We do not yet understand the nature of the process. The facts of proximity or remoteness with respect to other cultures and the ease of travel or communication certainly enter into the equation. Many of the elements of the inanimate and the biotic environment are emphatically relevant. The perception of local opportunity with respect to all manner of economic and social activity—a special place as glimpsed through the special lenses of the culture—would also appear to figure importantly. The regrettable fact that we cannot analyze all these many, still quite mysterious place-oriented interactions does not exorcize them. It is this zone of intersection between cultural process and the total character of places that is the special domain of the cultural geographer. In this volume, we seek to begin illuminating

9 We are much less interested here in mappable but adventitious spatial patterns. Thus one could plot the distribution in western and central Europe as of May 1945 of some millions of highly diversified displaced persons and prisoners of war or of the hundreds of thousands of American servicemen, tourists, students, and businessmen currently sojourning on foreign soil and be suffused with feelings of geographic virtue. Indeed the analysis of such maps could be scientifically interesting for a variety of purposes, but the contribution toward understanding the cultures of the migrant persons involved or of the areas in which they happened to be would be quite minor.

this zone as it exists in modern America, and to chart some of its bolder landmarks. But we cannot pretend that this is more than a timid beginning.

Population Mobility as a Factor in Cultural Change

The American population has been characterized by unusually great mobility from the very beginning of its history. As already noted, immigration brought different combinations of ethnic groups to different sections of the country, a process that helps account for the specific structure of early American culture. But spatial movements, on the part of regular migrants or the transient traveler, have continued vigorously past the initial step of immigration or pioneer settlement, and almost certainly have been accelerating or changing in kind, or both.[10] The individual migrant carries with him an appreciable portion, though never all, of the culture in which he has been nurtured. If he arrives as an ordinary single stranger in a community, the chances of his having any impact upon the total culture are infinitesimal. On the contrary, the likelihood is for his eventual assimilation into the host society, while contributing at best only a trivial trait or two: a single word, for example, or a card trick or anecdote. There is, of course, always the outside chance that, given just the right circumstances, major innovations can be implanted by a single newcomer or a very small group. Thus the almost mythical transfer of silk from China to Rome by a pair of monks or the introduction of textile machinery into New England by Samuel Slater, or the propagation of the Christian gospel into heathen territory by the solitary band of missionaries. But, generally, any appreciable modification of a regional culture calls for the arrival of a goodly number of strangers.

This process has obviously been going on quite actively among post-immigration American populations. Most conspicuous are the spatial transfers of ethnic and racial groups. The mass migration of the rural southern Negro into some Southern cities, but in much greater volume to the metropolises of the North, has appreciably changed the cultural character not only of the cities in question but of the migrant communities as well.[11] With their large, new, culturally vigorous Negro minorities or majorities, Washington, Detroit, Chicago, and Cleveland can never

[10] Wilbur Zelinsky, "The Hypothesis of the Mobility Transition," *Geographical Review,* 61, No. 2 (1971), 219–49.

[11] The literature on the urban sociology and demography of our Afro-American population has reached such staggering proportions that one must hesitate before unlatching the bibliographic floodgates. Perhaps the single most useful recent volume is Karl E. Taeuber and Alma F. Taeuber, *Negroes in Cities: Residential Segregation and Neighborhood Change* (Chicago: Aldine-Atherton, Inc., 1965); but Gunnar Myrdal, *An American Dilemma* (New York: Harper & Row, Publishers, 1944) will always remain indispensable. Within the geographic canon, the most useful statement is John Fraser Hart, "The Changing Distribution of the American Negro," *Annals of the Association of American Geographers,* 50, No. 3 (1960), 242–66. For a micro-geographic analysis of change in community and migrant, see the same author's "A Rural Retreat for Northern Negroes," *Geographical Review,* 50, No. 2 (1960), 147–68.

recapture their pre-World War II cultural personalities. In such places as Harlem and Atlanta, an Afro-American culture has come into being unlike any that existed earlier in the nation. There has been a rather less visible spatial re-sorting of other representatives of regional cultures: the heavy flow of Appalachian whites to Cincinnati, Detroit, and other midwestern communities has appreciably transformed them, for example. The eastward diaspora of the West Coast Japanese in the 1940's has made itself felt in the microgeography of culture in various Eastern metropolises. The recent migrations and recreational circulation of American Jews initially domiciled in the cities of the Northeast has resulted in marked Jewish influences on a number of rural tracts in the Northeast and cities in the West and South. Although it has never been methodically analyzed, there seems to be selective summertime migration of Czechs, Poles, Ukrainians, and probably other groups to certain rural retreats, always with some lasting effect. The spatial reshuffling of distinctive ethnic and regional groups, on either the interregional or intrametropolitan scale, may be more important than these few examples might suggest, for only the redistribution of the Negro population has attracted much scholarly attention.

Even less notice has been accorded the cultural repercussions of migrants selected in terms of characteristics other than ethnicity. As is the case with immigration from abroad, it is clear that patterns of internal migration do vary with age, sex, urban-rural residence, educational status, and class.[12] The evidence is much less accessible for migrational differentials among religious groups or occupations; but there is little doubt that they exist, are significant, and tend to produce new subcultural regional configurations. We may also speculate about correlations between propensity to migrate on the one hand and different types of temperament, levels of intelligence, political inclinations, and mental or physical health.[13] The possibilities for cultural evolution implicit in such movements are not limited to the simple concentration of similar individuals in certain areas or their weeding out from others. There is always the chance for some unanticipated reaction among newly juxtaposed cultural elements.

The primary motivation for migration into and within the United States has always been unquestionably economic. But from the beginning, the hope for monetary gain was supplemented strongly by other considerations. To take only the most obvious cultural facts, the decision by the potential European emigrant as to whether to try his fortune in

---

12 After more than three decades, the best general statement on the topic is still Dorothy Swaine Thomas, *Research Memorandum on Migration Differentials,* Bulletin 43 (New York: Social Science Research Council, 1938).

13 Some tentative findings are, in fact, available for the impact of mental disease upon the propensity to migrate. See Judith Lazarus, Ben Z. Locke, and Dorothy Swaine Thomas, "Migration Differentials in Mental Disease: State Patterns in First Admissions to Mental Hospitals for All Disorders and for Schizophrenia, New York, Ohio, and California As of 1950," *Milbank Memorial Fund Quarterly,* 51 (January 1963), 25–42.

America was closely related to ethnic, religious, and political identity. In the contemporary United States, with an exceptionally high spatial mobility and a general affluence that may be undercutting the primacy of the economic motive, it is logical to expect many persons to seek or shun specific places because of their personal cultural predilections. The implications of such a trend are examined in chapter 4.

The Diffusion of Innovations

The inducement of social, economic, and cultural change through the spatial diffusion of innovations has recently and justifiably received much attention from students of geography and other fields.[14] If this interest was initially centered on the outward propagation of various forms of technology on a macro scale by anthropogeographers and anthropologists,[15] the more recent focus has been upon the finer locational detail. A series of rigorous studies of the mechanics of contagious diffusion among neighboring individuals, the "telling" of unfamiliar information among pairs of persons, has led to clarification of what appears to be a stochastic process. Indeed, real-life situations can be simulated through Monte Carlo methods.[16] Insofar as such analysis is applicable to cultural geography, it carries with it the assumption that the total culture consists of a bundle of traits, a collection of particulate objects. This is at best a half-truth, but, up to a point, a quite useful one. Although the Hägerstrandian diffusion of innovations through personal contact among essentially stationary actors commends itself to the geographic analyst because it can be handled with relative methodological elegance and rigor, it is only one of three mechanisms whereby novel ideas can be spread from place to place: these are contagious diffusion; migration; and telecommunications. Under typical American conditions, the first of these may be the least important.

The movement of individuals from one cultural milieu into another also automatically means the transfer of their culture, at least within their minds. Whether there is effective transposition to the host population of any of the imported traits depends upon many variables. How

[14] The seminal work in this area is Torsten Hägerstrand, *Innovation Diffusion as a Spatial Process* (Chicago: University of Chicago Press, 1967); and for developments since 1953 (the date of original publication), the "Postscript" (pp. 299–324) by Allan Pred, the translator and editor of the volume, is quite useful, as is Ronald Abler, John S. Adams, and Peter Gould, *Spatial Organization: The Geographer's View of the World* (Englewood Cliffs, N.J., Prentice-Hall, Inc., 1971), pp. 389–451.

[15] Notable examples of the treatment of this theme by anthropogeographers at the macro scale include Carl O. Sauer, *Agricultural Origins and Dispersals* (New York: American Geographical Society, 1952) and a number of case studies scattered through Joseph E. Spencer and William L. Thomas, Jr., *Cultural Geography; an Evolutionary Introduction to Our Humanized Earth* (New York: John Wiley & Sons, Inc., 1969). See also George F. Carter, *Man and the Land: A Cultural Geography,* 2d ed. (New York: Holt, Rinehart & Winston, Inc., 1968).

[16] As elegantly exemplified in Richard Morrill, "The Negro Ghetto: Problems and Alternatives," *Geographical Review,* 55, No. 3 (1965), 339–61.

many migrant carriers are involved? How permanent is the migration, or how prolonged the period of contact? What sorts of transactions occur between newcomers and older residents? What is their relative status? In addition to regular migrants, there are transients, for example, tourists, businessmen, students, servicemen, itinerant workers, conventioneers, and others. Such movement is particularly vigorous in the United States and further lubricates the cultural change already stimulated by large-scale migration.

The existence of several media for rapid, even instantaneous long-distance transmission of information adds a third major dimension to the phenomenon of diffusion. The early establishment of an effective American postal system meant the circulation of personal correspondence and printed newspapers, periodicals, and books to all corners of the land within a few days or weeks. The post was followed by telegraph and cable by the mid-nineteenth century; and later by telephone, tele-type, mass-produced phonograph records and tapes, radio, and television. During the Paleolithic period, it might take millenia for a welcome inno-vation to reach all prospective clients, if indeed it ever did realize its outermost territorial range. By early modern times, the pace had quick-ened noticeably. It was then a matter of only decades or years, so that the acceptance of tobacco throughout the Old World was consummated in less than a century.[17] Today the appetite for novel artifacts can be excited literally overnight; and only problems of manufacture or delivery inhibit instant national or worldwide adoption.[18]

The veritable revolution in communications has been so profound and so sudden that its geographic implications have not been appreciated or seriously studied. Thus contagious diffusion, the dominant premodern mode of effecting cultural change spatially, may be giving way to migra-tion and telecommunication, as new ideas and people vault rapidly over great distances and the spatial patterning of acceptors may be governed more by their ranking within one or more hierarchical systems than by propinquity alone. Thus the size or status of a city, college, or firm, or the standing of an individual in terms of social class, occupation, educa-tional attainment, or ethnic or racial group may be decisive as to when an innovation is adopted.

If we turn to the earlier history of the United States, it would appear that the dissemination of innovations by mobile individuals was at least as effective as contagious diffusion in modifying the cultural map. Such studies as we have lead to two conclusions: (1) the most critical form of cultural diffusion was carried on by those pioneer settlers who estab-

17 Louis Seig, "The Spread of Tobacco: A Study in Cultural Diffusion," *Pro-fessional Geographer*, 15, No. 1 (1963), 17–21.

18 I can attest to this phenomenon from personal experience. In August, 1958, I drove from Santa Monica, California to Detroit at an average rate of about 400 miles per day; and display windows in almost every drugstore and variety store along the way were being hastily stocked with hula hoops just off the delivery trucks from Southern California. A national television program the week before had roused instant cravings. It was an eerie sensation, surfing along a pseudo-innovation wave.

lished the first effective settlements in various parts of the country; and (2) there were only a very few centers of especially active cultural ferment, of invention and elaboration, which acted as culture hearths, or in some cases funnels, from which local or exotic innovations spread over extensive areas during or after the period of frontier settlement (Figure 3.2).[19]

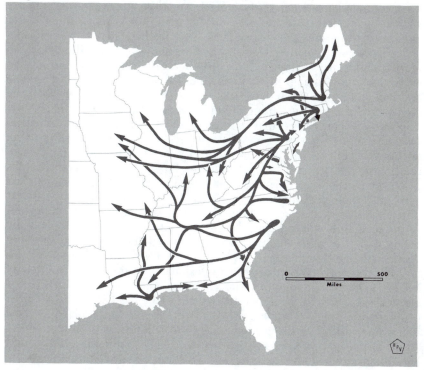

FIG. 3.2   *The movement of ideas in the Eastern United States (after H. Glassie).*

Care must be taken to avoid misleading analogies with other modes of spatial movement in discussing the diffusion of specifically cultural items. Although useful parallels can be drawn, and certain common mathematical formulations may apply to them all, the physical dispersion of gaseous or liquid substances, the spread of an epidemic, and the conquest of territory by a new plant or animal species do differ in fundamental ways from the spatial propagation of new practices among human societies. Furthermore, the diffusion of technological or economic inno-

[19] A marked localization of inventions and innovations in late nineteenth century America is indicated in an analysis of the location of patent applications in Allan R. Pred, *The Spatial Dynamics of U.S. Urban-Industrial Growth, 1800–1914: Interpretive and Theoretical Essays* (Cambridge, Mass.: M.I.T. Press, 1966), pp. 86–134.

vations is not quite equivalent to the processes involved in implanting novel cultural traits in alien soil. With the former, there are objective standards of performance, even though, of course, the ultimate definition of value is subjective. Competitive items can usually be readily ranked, so that if the freshly invented doodad can do the job better, it will be accepted as quickly as money and opportunity permit, whereas its predecessor is unceremoniously junked. Thus the stone axe never had a chance once metal tools appeared on the scene; the modern water closet has swept all before it; if you wish to see a leather tankard being made, you will have to visit Williamsburg, Virginia; and such things as the bicycle and plastic raingear have penetrated almost every part of the world, including some primitive societies otherwise adamantly set against any sort of change. Their virtues of low cost, compactness, or efficiency are just too compelling. Strong cultural barriers may slow down, but not stop, new technologies that are demonstrably superior. Thus many societies can fiercely resist improved contraceptive devices or better forms of nutrition, but usually such resistance is a rear guard action. Witness the growing acceptance of milk among Japanese youngsters who are being convinced that it is good for them, whatever their grandparents may mutter about the barbarous habit.

When purely cultural preferences are at issue, it is much more difficult to scale utility or to predict rates of acceptance. Is the miniskirt really better than the ankle-length gown? (And will either linger for more than a few months or years at a time?) Is abstract expressionism inferior to op or pop art? Why does the Englishman continue to prefer tea to coffee? Why should one cultivate the twist, the frug, or whatever the current craze might be rather than the foxtrot or the waltz? Is pizza really that much better than whatever edibles it has replaced in the American cuisine? Is the contemporary ranch house truly preferable to a turn-of-the-century Queen Anne house? All these questions and countless others are matters of taste and "fashion" controlled by unknown cultural-historical principles;[20] and there is no "rational" set of scales to measure whether choices among cultural options result in improvements.

We need not narrate in detail here the filling in of the American territory by pioneer settlers, with their cultural baggage. The general story has been well told, in word and map, in a number of publications,[21]

---

[20] The study of fashion is evidently still regarded as too frivolous by social scientists. Virtually the only serious general essay on its sociology is Herbert Blumer, "Fashion: From Class Differentiation to Collective Selection," *Sociological Quarterly* (Summer 1969), pp. 279–91. That it can prove to be a productive line of inquiry is amply demonstrated in J. Richardson and A. L. Kroeber, "Three Centuries of Women's Dress Fashions: A Quantitative Analysis," *Anthropological Records*, 5 (1947), 111–53.

[21] For example, Ralph H. Brown, *Historical Geography of the United States*, (New York: Harcourt Brace Jovanovich, Inc., 1948); Ray A. Billington, *Westward Expansion: A History of the American Frontier*, 3d ed. (New York: The Macmillan Company, 1967); and Randall D. Sale and Edwin D. Karn, *American Expansion: A Book of Maps* (Homewood, Ill.: Dorsey Press, 1962).

and the specifically cultural dimension of frontier advance is noted in the following chapter. The essential geometry of the taking of the American land is that of a frontier zone irregularly accelerating westward. This zone spread out from three nuclear nodes along the Atlantic Seaboard: southern New England; the Midland, based in the Delaware and Susquehanna valleys; and Chesapeake Bay. A fourth source, the French-American establishment along the lower St. Lawrence, proved to be of only ephemeral importance. Similarly, the various Hispanic-American and aboriginal groups swallowed by the dynamic Anglo-American frontier were minor sources of influence for the national culture. There has been some (two-way) traffic in cultural effects across the Mexican and Canadian borders; but most of the significant innovations have moved internally within and among American regions, or been transferred by specific regions through which overseas items are imported and spread inland.

Despite the stubborn romantic inclination to believe otherwise, the settlement frontier cannot be credited with the origination of any important inventions, material or otherwise. Even the new gear most closely associated with frontier life (for example, the Conestoga wagon,[22] the Colt pistol, the Springfield rifle) came from the "effete" East. But one cultural phenomenon, the unconscious rediscovery of a primitive or pan-European cultural substratum common to the varied ethnic strains that streamed across the Atlantic, is of more than passing interest. A groping backward into a half-forgotten past was resorted to by rusticated groups suffering temporary physical and social deprivation. We find evidence for it, for example, in the improvising of folk housing of a Ur-European character[23] and the resuscitation of some very early motifs in gravestone carving.[24] Such resurrection of the past is meaningful insofar as the frontier still persists physically in many a backwater tract, or ethereally as a strong mythic force in American thought. In general, however, the frontier played a passive role in the American cultural drama, as the remote destination for things, not as their source.

During the seventeenth century and much of the eighteenth century, Tidewater Virginia was preeminent in receiving fresh European ideas and disseminating them, if not in inventing new ones. This position is well indicated by such physical evidence as architecture and furniture;[25] and it is not unexpected in the light of the area's economic and political strength. Other early Southern coastal zones, such as those centered around Charleston, Savannah, and New Orleans, were much less influential at the national scale, largely because of weaker populations and

[22] Gary S. Dunbar, "Wagons West—and East!" *Geographical Review,* 55, No. 2 (1965), 282–83.

[23] Alan Gowans, *Images of American Living: Four Centuries of Architecture and Furniture as Cultural Expression* (Philadelphia and New York: J. B. Lippincott Co., 1964), p. 60.

[24] This notion is developed in some detail in Allan I. Ludwig, *Graven Images: New England Stonecarving and its Symbols, 1650–1815* (Middletown, Conn.: Wesleyan University Press, 1966).

[25] Gowans, *Images of American Living,* pp. 173–222.

economies and poorly developed connections with the continental interior.

By the end of the eighteenth century, cultural leadership had passed on to the more developed portions of New England, then rapidly ascendant in commercial and manufacturing activity. Throughout the nineteenth century, this region was clearly the most fecund and powerful, setting an example for the whole country in many departments of higher human endeavor. By virtue of the vigorous out-migration of New Englanders and the diffusion of ideas and objects through other means, the national patterns in industrial technology and mechanical devices of all sorts, higher education, science, literature and the other fine arts, theology, political ideas, manners, and domestic and public architecture were largely controlled by this single small region.[26] In fact, the cultural geography of nineteenth century America can be described, without serious exaggeration, as the continual pumping and spraying outward to west and south of a great array of novelties, locally invented or imported from abroad, from the New England reservoir. As New England folk and ways of life moved westward into upper New York State, there to intermingle fruitfully with other streams of settlement and ideas, a subsidiary culture hearth germinated within "New England Extended," especially within the "Burned-Over District" of west central New York.[27] What little work we do have on the cultural geography of the tract strongly suggests its potency in inventing and disseminating a variety of religious and social items.

How great a role the Midland culture area, centered in eastern Pennsylvania, may have played as a source of nationally significant items is much less clear. There was certainly some transmission of agricultural practices, religious denominations, and architectural forms (for example, barns and log construction) among other things;[28] but it is possible that the Midland's total contribution to the national cultural pattern may be less than would be suggested by its strategic location and early prosperity.

With the achievement of major national strength and an enormous settled territory by the mid-nineteenth century, the simple geometry of colonial cultural change and diffusion became decreasingly relevant. In place of a small set of Atlantic Seaboard metropolises and their immediate hinterlands, which conducted European cultural commodities

26 Lois (Kimball) Mathews Rosenberry, *The Expansion of New England: The Spread of New England Settlement and Institutions to the Mississippi River, 1620–1865* (Boston: The Houghton Mifflin Company, 1909); and Stewart H. Holbrook, *The Yankee Exodus: An Account of Migration from New England* (New York: The Macmillan Company, 1950).

27 Whitney R. Cross, *The Burned-Over District: The Social and Intellectual History of Enthusiastic Religion in Western New York, 1800–1850* (Ithaca, N.Y.: Cornell University Press, 1950).

28 Carl O. Sauer, "The Settlement of the Humid East," *Climate and Man: Yearbook of Agriculture 1941* (Washington: U.S. Department of Agriculture, 1941), pp. 164–66; Dunbar, "Wagons West—and East!"; Harold R. Shurtleff, *The Log Cabin Myth* (Cambridge, Mass.: Harvard University Press, 1939); and Edward T. Price, "The Central Courthouse Square in the American County Seat," *Geographical Review*, 58, No. 1 (1968), 29–60.

into North America or devised new ones, two other processes began to take over: the scattering of inventiveness, and an obvious cultural convergence. Welcome mechanical and social inventions could appear almost anywhere, except perhaps the remoter settlement frontiers, within a rapidly maturing, ever more self-assured society. Thus bourbon whiskey first appeared in Kentucky; an assortment of agricultural devices emanated from California and the Middle West; women's suffrage won its first major victory in Wyoming; the more daring experiments in educational practice were hatched on Middle Western campuses; and by 1900 Chicago may have become the most exciting place in the country for the writer or architect. If there has been a discernible reordering of peaks and valleys on the American map of relative cultural vigor over the past two centuries, the altitudes would now seem to be greatest in the Far West and Southwest. The Megalopolitan zone of the Northeast, most particularly Greater New York City and southern New England, may still originate and broadcast many ideas of note, but increasingly the newer modes of behavior and thought, as well as inventions of all sorts, seem to germinate most often in the metropolitan zones of California and neighboring states. However, the cultural map of late twentieth century America is not just the mirror image of the early nineteenth century. No longer is the simple physical flow of ideas and people through two-dimensional space the dominant process. The rules of the game have undergone profound, but poorly understood, revisions, so that the current mechanisms of change, thanks largely to radically different modes of communication and social structure, may bear scant resemblance to those of the past. One of the more obvious symptoms of the changing forms of change is the apparent homogenization of the nation, and indeed much of the world, through cultural convergence on a grand scale.

CONVERGENCE VS. DIVERGENCE. For the first 200 years of the European presence in North America, the forces of cultural diversification seem to have prevailed. By virtue of the differential migration over time and space of varied peoples and traits from overseas, the interaction of the newcomers with novel habitats and neighbors, and the spontaneous evolution of relatively isolated communities, the regions of Anglo-America diverged ever further from the parental European cultures, and each locality tended to become less similar to all others. Then, roughly during the period 1830–1850, a reversal of trend seems to have begun, as a set of powerful processes fostering national uniformity made themselves felt. A continental network of highways, canals, railroads, postal routes, and telegraph lines began to materialize, along with truly national business corporations. Mass production and mass consumption of anonymously made commodities began to edge out local artisans making individualized things for small markets.[29] Tastes and ideals were standardized with the proliferation of nationally distributed magazines, newspapers, books, popular songs and advertising, the itinerant dramatic troupe,

29 John A. Kouwenhoven, *Made in America: The Arts in Modern Civilization* (Garden City, N.Y.: Doubleday & Co., Inc., 1948).

circus, and lecturer, and the great popularity of school textbooks (in vast editions), carpenter's and other trade manuals, manufacturers' catalogs, fashion books for ladies, and eventually mail-order catalogs. All these trends have, of course, accelerated in recent decades.

And they have gone well beyond our national boundaries. Better and faster methods for dispatching information and people hastened the acceptance of new trends in European technology, thought, and arts.[30] They also made possible a rapid, reverse flow of native American products, not only to Europe but to much of the non-American world. The cargo list of exports includes not only American variants of standard agricultural or industrial items (for example, motor vehicles, fabrics, and electrical goods), but many things that are undeniably American: chewing gum, basketball, comic strips, cornflakes, cola beverages, the major pesticides, jazz and its recent derivatives, casual clothing, the dominant forms of cinema and television entertainment, the supermarket, the drivein, the motel, Mormons and Jehovah's Witnesses, the Western genre in print and film, facial tissues, ballpoint pens, and hundreds of words, American in origin but now universal in currency.

To allay somewhat the accusation of foreigners that Americans seek to coca-colonize the world, we have taken as avidly as we have given. Few nations have become as cosmopolitan in consumer tastes in recent years as has the United States, for reasons going beyond our large, polyglot immigrant population. The menus of our restaurants and the shelves of our groceries are stuffed with delicacies from scores of exotic lands. Popular music has been deeply tinged by the styles indigenous to Brazil, Polynesia, Mexico, India, Israel, and a dozen other alien nations. Many forms of literature, music, painting, sculpture, architecture, theatre, and dance have been thoroughly internationalized. The American cognoscenti covet Eskimo carvings, African masks, Chinese ceramics, and performances on the koto, sitar, and bazouki. The ever-volatile styles of women's (and more recently men's) dress, jewelry, and coiffure appear to reflect every possible variation upon ethnic themes.

Evidence for the impact of modern technology and marketing procedures in smoothing out cultural irregularities is too abundant and blatant to be ignored. Within the United States, one suburban subdivision or shopping center may be almost indistinguishable from another. Motels, filling stations, magazine racks, drug stores, high schools, radio program formats, chain restaurants, and the seemingly unlimited array of franchise retail enterprises are all very much peas in a pod, whatever the city or state. Every place has become Anyplace. At the international scale, the casual visitor cannot tell one country from another by style of airport, hotel, airline, or apartment building.

But appearances can be deceiving. In trying to measure degrees of similarity among different places in a rigorous fashion, one must answer two basic questions: the precise specification of what it is that is being measured; and the application of suitable scales of measurement. For certain aspects of human activity, for example, demographic and eco-

---

30 Spencer and Thomas, *Cultural Geography*, 277–326.

nomic characteristics, generally acceptable definitions and modes of measurement are available. It has also been demonstrated rather convincingly that convergence has occurred in recent decades in both sets of characteristics within and among highly advanced nations (at least when populations are aggregated within sizeable reporting units).[31] In the cultural realm, and especially with sociofacts and mentifacts, the technical problems are much more severe. As noted earlier, cultural systems are exceedingly complex, many-dimensional entities. It is difficult to decide which dimensions particularly merit notice, what relative weights should be assigned to them, and how they might be measured; and the optimal mathematical procedures for comparing the resulting values as between places through space and time have yet to be specified.

Apparently only two serious attempts have been made to test the degree of cultural convergence among different sections of the United States. The first of these, using the results of various public opinion polls over a recent 30-year period, arranged by region and other variables, failed to disclose any convergence, and indeed suggested the opposite trend.[32] But some basic, intractable problems in interpreting the data suggest caution in accepting these results. In the other study, involving an analysis of choice of given names for males in selected counties in the eastern United States in 1790 and 1968,[33] a substantial amount of convergence, and an apparent erosion of earlier culture areas, was demonstrated, but again with some equivocation. The array of names in the later year was immensely greater and richer than in eighteenth century America; and it is entirely possible that aggregating data at the county level may have masked much within-county—or rather within-society—diversity in naming complexes. Paradoxically, the answer to the question of whether the United States (or the entire modernized world) has been becoming more homogeneous may be: "yes and no."

Within the familiar two-dimensional space used by geographers in their cartographic analyses, cultural distances among ever more complicated entities would seem to be shrinking; but modern man, torn loose from conventional bounds of place or social and biological descent, may well be feeling his way into a number of newly discovered dimensions. The opportunities for personal choice, more complete individuation, and the formation of new social and cultural entities may have been greatly enhanced. Thus although most places may have begun to look alike, in important ways not usually susceptible to casual visual observation they may have started down fundamentally different routes. In sharp contrast,

[31] Hope T. Eldridge and Dorothy S. Thomas, *Population Redistribution and Economic Growth, United States, 1870–1950, III. Demographic Analyses and Interrelations* (Philadelphia: American Philosophical Society, 1964); and S. Labovitz, "Territorial Differentiation and Societal Change," *Pacific Sociological Review,* 8 (1965), 70–75.

[32] Norval D. Glenn and J. L. Simmons, "Are Regional Cultural Differences Diminishing?" *Public Opinion Quarterly,* 31 (1967), 176–93.

[33] Wilbur Zelinsky, "Cultural Variation in Personal Name Patterns in the Eastern United States," *Annals of the Association of American Geographers,* 60, No. 4 (1970), 743–69.

the communities of the premodern past may have displayed the greatest imaginable superficial differences, but the most striking isomorphisms are revealed to the persistent analyst. The essential rules for society and culture were repeated in place after place, varied superficially but possessing the same components connected in similar ways.

This rather esoteric argument can be condensed into the following summary: the most advanced nations of the world, notably the United States, have entered an era of truly revolutionary social and cultural transformation such that the inward structure of society and culture may be skipping past a phase boundary into a totally new set of conditions. The results, insofar as they can be analyzed geographically, on maps or otherwise, are so nonconformable with the products of earlier situations that comparisons are awkward or impossible to make. To offer an analogy, the problem is rather like comparing the territorial range and activity patterns of birds with those of their earthbound reptilian ancestors.

The American House

Many of the generalizations on spatial processes in the preceding pages, along with those on the origins and special identity of American culture, can be fleshed out by considering two very different sets of phenomena: the American house and patterns of church membership.

We are concerned here with the dwellings of the great majority of the American population—what might be fairly characterized as "folk housing"—and buildings that never enjoyed the professional attention of architects.[34] Although basically an artifact and one serving some urgent physical functions, the house is also the product of a complex set of societal and psychological factors, all filtered through the sediments of history. It is really as much sociofact or mentifact as artifact. When fully interpreted, the form and uses of the house tell us much, not only about the physical locale and the technology of the place and era, but also about the source and dates of the builder or renovator, the contacts and influences he experienced, his ethnic affiliation, and possibly also class, occupation, and religion.[35] In a very real sense, the house is the family's universe in microcosm, the distillation of past experience and a miniature model of how it perceives the outer world, as it is or perhaps even more as it should be.

In essence, the American house is a European import. Or, rather,

---

[34] Geographers who are guilt-ridden in the knowledge that no set of phenomena is their exclusive disciplinary property may draw some solace from folk housing. It is an orphan topic in the United States, one neglected by architects and anthropologists, and given serious attention only by a few eccentric geographers.

[35] The tangible meanings conveyed by house morphology are probed in Amos Rapoport, *House Form and Culture* (Englewood Cliffs, N.J.: Prentice-Hall, Inc., 1969); and Peirce F. Lewis, "The Geography of Old Houses," *Earth and Mineral Sciences* (Pennsylvania State University), 39 (February 1970), 33–37. Deeper layers of significance are charted in James Agee, *Let Us Now Praise Famous Men* (Boston: The Houghton Mifflin Company, 1941); and Gaston Bachelard, *The Poetics of Space* (New York: Orion Press, 1964).

it is a uniquely new object reconstituted from a number of earlier European fragments, which then evolved in a special way in accordance with the peculiarities of American life. Ideas, usually subconscious, as to the proper way to construct a dwelling varied from place to place along the colonial Atlantic Seaboard, depending upon time, sources, and conditions of colonization. Quite early, distinct regionalisms in house styles began to develop. But equally early, the designs, and often some of the materials, of the homes for the wealthy were imported intact from northwest Europe. It was only about the time of American Independence that an indigenous professionalism in architecture for homes and public structures began to develop. Yet, if the homes of the common people of Massachusetts, the Hudson Valley, and North Carolina were all distinct from the beginning, they shared nevertheless some unmistakably American traits. These, in turn, reflect the primordial notions about house morphology underlying folk architecture throughout a good part of Europe. These could surface only under the relatively primitive conditions of a new North American society. There was little borrowing of aboriginal building techniques, and then only locally and temporarily. And no African heritage can be discerned among the structures built by or for the slaves, except perhaps in the still unstudied rural Negro churches.

Each of the three principal colonial culture hearths—southern New England, the Midland, and the Chesapeake Bay area—developed its own set of house types and other sorts of buildings at an early date, and so also did a number of minor subregions. We can trace the westward thrust of settlers and ideas into the continental interior from these seedbeds of American style quite precisely in the field, at least up until the mid-nineteenth century, by plotting the location of surviving examples of these regional types.[36] And we can also observe the ways in which various strands of culture flowed together and sometimes produced new regional blends. In fact, the quasi-archaeological technique of studying older dwellings is one of the better, if more laborious, ways of charting the past or present extent of culture areas (or the microgeography of older cities) and of gaining insight into the historical geography of American ideas.

As in other departments of cultural practice, there is a striking depletion of individuality and inventiveness in building styles and ornamentation as one moves away from the early communities in the East to the relatively accessible regions of the West. In part this was presumably the result of relative isolation during formative years. In place of the truly riotous exuberance of form within a single long-lived New England village or even a single block in an eastern Pennsylvania

[36] Fred Kniffen, "Folk Housing: Key to Diffusion," *Annals of the Association of American Geographers*, 55, No. 4 (1965), 549–77; Wilbur Zelinsky, "Where the South Begins: The Northern Limit of the Cis-Appalachian South in Terms of Settlement Landscape," *Social Forces*, 30 (1951), 172–78; Henry Glassie, *Pattern in the Material Folk Culture of the Eastern United States* (Philadelphia: University of Pennsylvania Press, 1968); and Lewis, "The Geography of Old Houses."

borough—a variety, by the way, still somehow harmonious—there is the monotony and stunted imagination of the Middle Western or far Western residential neighborhood. This tendency is even more marked within business districts. An attenuation of style is observable even within so simple a category as techniques of log-house construction; and in a study of Georgia examples, a striking contrast was seen between the earlier, richer repertory of log-house forms in the (older) North and the later, stripped-down set found toward the (younger) South.[37]

The structures built after about 1850 tell quite a different story from their predecessors. By that date there had been sufficient mingling and hybridization of the original colonial styles within the great central expanses of the country to produce a recognizably national group of building types; and the pervasive new modes of communication and manufacturing had started to iron out regional departures from mass norms. For the past century or so, domestic building styles are closely correlated with date, and are much more sensitive to rate of diffusion down a social or cultural hierarchy than to territorial location. Yet, however standardized American building practices may be becoming, the house, old or new, is still packed densely with information about the national ethos, the dealings between man and habitat, and the changing configurations of our cultural geography. Surely the geographer must concern himself with the form and meaning of the objects he studies if any real sense is to be made of their spatial array or of their processual linkages with other phenomena in space and time.

Several attributes of the American house, past and present, bespeak important peculiarities of the national character. Perhaps the most obvious is the lavish use of space, both in the sheer size of the house proper and in the largeness of the residential lot. Furthermore, a disproportionately small fraction of the population live in apartment buildings or other multiunit structures; many of those who do are college students, convicts, the indigent or ailing elderly, and other institutionalized populations.[38] Except in the most congested of urban settings, as in New York City, Boston, or San Francisco, the one-family dwelling is a free-standing unit, with at least a token patch of space between it and its neighbors. Row housing is a phenomenon restricted to urban neighborhoods in the Northeast that were built about two centuries ago, apparently in imitation of Northwest European models.

Although much of this expansiveness might be explained away, the residual cultural factor bulks large. Land was cheap and abundant, and still is relative to land in most parts of the world; but the urge toward very large, isolated, individual properties seems to go beyond any rational economic reckoning. This is most obvious in the isolated farmstead,

37 Wilbur Zelinsky, "The Log House in Georgia," *Geographical Review*, 43, No. 2 (1953), 173–93.

38 This aversion to apartment life may derive, in part, from British attitudes. Although statistics on housing characteristics are designed by various national governments so as to make valid international comparisons almost impossible, simple observation leads one to believe that the urban English are much fonder of the single-family dwelling than are any of their European neighbors.

for which no convincing argument can be made in terms of transportation systems or social utility. If the propensity of Americans toward larger lots is being more fully realized now with growing affluence and the lateral spread of cities, the cubic volume of the structure has been decreasing. This is largely because of rising costs of materials and labor, the scarcity of servants, and the shrinkage in average size of household. But American homes are still bulky by any universal criteria; and those of the nineteenth century middle class were often of incredible proportions, even after making all allowances for number of hired hands, children, and other kinfolk, or the provision of closets and storerooms for the accumulations of a super-productive economy. One cannot help but speculate that these dimensions reflect an optimistic, aggressively extroverted view of the world and the American's place in it.

The fact that Americans may be profligate with space but niggardly with time also appears in their building technology. Great store is set upon quickness of construction, and it is not by chance that the United States has originated or perfected most leading methods of building prefabrication, or that the technique of balloon frame construction, one that reduced costs and workdays for wooden edifices so markedly, won such instant, universal acceptance.[39] The commercial hotel as we now know it and its up-to-date offspring, the motel, both of which were nurtured in, and are ubiquitous throughout, North America, also embody the themes of haste and transience.[40] Note also the emphasis on the garage, often an integral, conspicuous part of the house, sometimes even threatening to dominate the nonautomotive segment of the structure. The stress on transience, clearly evident in the early shelters of pioneer settlers, has waxed rather than waned in recent decades. Few Americans build a house with the intention of occupying it for a lifetime or passing it on to their children;[41] indeed our population is so mobile that every other family changes its abode every decade. These urges toward transience and mobility receive their ideal embodiment in that superlatively American invention, the house trailer, which is virtually nonexistent outside North America. Nomadism as a way of life, though not yet well documented by the social scientist, may have made marked progress during the 1960's among some members of the so-called "counterculture," for whom the VW microbus or some other such vehicle in nearly constant motion may be the only true home. But even among the most respectable strata of American society, transience is revered. Witness the recent popularity of high-rise office and apartment buildings designed to obsolesce and be razed after a few years, but only after the maximum tax advantage has been squeezed out of them.

Partly because the ordinary American house is so transitory a phe-

[39] Daniel J. Boorstin, *The Americans: The National Experience* (New York: Random House, Inc., 1965), 148–52.

[40] Boorstin, *The Americans*, 134–47.

[41] Such heirlooms are certifiable curiosities. In Michigan, farmhouses that have been continuously occupied by members of the same family for 100 years or more are designated "Centennial Homes" by a state agency, and are so identified for the passerby by means of a metal plaque.

nomenon, but basically because of a fundamental restlessness of character, we have witnessed a dizzying procession of building styles and fads, one following hard upon the heels of another. As already noted, the modern American house (or commercial or public building) is a much better indicator of date than of locality. So avid is the appetite for novelty that the architect has ransacked virtually every historic era and most regions of the world in search of inspiration.

A cheerful extroversion of personality is writ large in the American house and its surroundings, which do double service as status symbols as well as shelters. The penchant for large glass windows, a trend that shows no sign of abating, goes beyond a normal craving for natural illumination and creates some serious problems in heating and upkeep. With the advent of the picture window craze, it becomes especially clear that the house is designed to serve as a display case, to advertise to the world at large the opulence and amiability of the household. The same outward-going personality is visible in the extravagant development of the porch. Although the ultimate origins of the porch (or portico or piazza) are obscure—the British may have hit upon the idea in the West Indies or India—no other national group has seized upon the device with such enthusiasm. During its apogee around 1900, few self-respecting American houses were without one, and many houses were encased with a porch on two or three sides and possibly on the second as well as the ground level. These open-air extensions of the house—literally a perpetual "open house"—were the stages upon which much of the social life of the family was enacted during the warmer seasons. This obliteration of the distinction between inside and outside, totally at variance with Northwest European antecedents, is carried even further in much avant-garde architecture, especially in the Pacific Coast states.

The same impulses that favored the efflorescence of window and porch seem to lie behind the almost pathological fervor with which grass lawns are tended—and front fences or hedges are frowned upon. (Do we have here the democratization of the British baronial estate?) There is still much to be learned about the culture of a group, including the American, through the microgeographic analysis of their house gardens.[42] However, one peculiarity of American landscaping leaps to the eye: contrary to general usage, gardens are not invariably private spaces behind walls or hedges, but are often aggressively public, placed on the street side of the house. And since the lawn itself is basically ornamental or symbolic, intended much more for show than for any sort of play or foot traffic, it must be considered along with shrubs and flowers as a badge of membership in a cheerful, outgoing democratic society, but one in which the privileges of a powerful individualism must also be made manifest. The lawn is also significant as a shorthand symbol for the edenic ideal that is so strong an undercurrent in American thought. It is also appropriate to indicate here a rather cavalier disregard of

[42] Writings on Oriental gardens provide models that might well be emulated by American scholars. Some recent items are noted in Yi-Fu Tuan, "Man and Nature," *Landscape*, 15, No. 3 (1966), 32.

environmental conditions on the part of designer and homeowner, an attitude stemming from an overriding self-confidence and material abundance as much as from ignorance. Except for a tiny minority of interesting exceptions, ideas are *imposed* upon the land, however inappropriately (for example, lawns on the sands of Florida or picture windows in the subarctic). Climate, slope, drainage, geology, soil, and natural vegetation seem to matter little.

Paradoxically, despite its openness, the American house also attests to the supremacy of the private individual. We have already noted the aversion to inhabiting multifamily structures and the impulse to create token open spaces between neighbors. But it is in its internal arrangements, with the great stress upon isolation, the multiplicity of doors and closed spaces, and the segregation of specific functions, that the pervasive American privatism comes fully to the fore.[43] In addition, taking the house and grounds as a single entity, there is the starkest kind of contrast between the American's attitude toward his private bubble of space and that toward all public spaces. All self-respecting householders spend an inordinate amount of time caring for yard and garden and on keeping the interior as antiseptic and spotless as human ingenuity can manage. But public spaces, including sidewalks, thoroughfares, roadsides, public vehicles, parks, and many public buildings reveal a studied neglect and frequently such downright squalor that it is difficult to believe one is encountering a civilized community.

Finally, the American house quite neatly illustrates an all-powerful mechanistic vision of the world, for it is a carefully manipulated machine, a working model of what Americans feel the cosmos fundamentally is— or could be induced to be. The internal physiology of the American house represents a truly awe-inspiring triumph of the mechanical arts. The list of inventions promoting domestic comfort and convenience attributable to American ingenuity is long and fascinating. Elaborate and ultimately effortless central heating (and cooling) systems have been devised; then, like the other wonders noted below, made available to the world at large.[44] The water supply has been brought indoors, and thoroughly rationalized. American plumbing is the eighth wonder of the world, and the Great American Bathroom a veritable glittering cathedral of cleanliness. The kitchen is also a marvel of efficiency and clever design, incorporating a multitude of American "firsts." Similarly advanced are lighting, electrical wiring, laundry systems, and rubbish disposal. As already intimated, the layout of the house is thoroughly programmed, with a specific function, and usually no other, designated for each space. Thus we have carried to its logical extreme that spatial separation of place of work and place of residence that is so distinctive and important a feature of the larger landscape. This statement also applies to the farmer who

---

[43] The contrast between the internal plans of American and Chinese houses in this respect is almost total. See Francis L. K. Hsu, *Americans and Chinese: Two Ways of Life* (New York: Abelard Schuman, Inc., 1953), 68–70.

[44] We can forgive the British many things, but never their barbaric indifference to the virtues of central heating.

resides at his work place but works in different buildings on or near the farmstead from that in which he eats and sleeps. In almost every respect, then, the American house is a completely appropriate capsule world, fleshing out the main principles, myths, and values of the larger cultural system.

Geography of Denominational Membership

If the form, substance, and spatial patterns of American houses convey much vital information about the totality of the national culture, a great deal can also be decoded from the data on religious affiliation, but from a rather different perspective.[45] The trend in house morphology seems to be decidedly toward place-to-place uniformity; but, in terms of religion, the United States has been a land of phenomenal diversity from the very beginning, and this heterogeneity fails to show the slightest sign of slackening. In fact, no other nation begins to approach the United States in this respect. The best account available lists some 251 religious bodies able and willing to offer statistics on membership, but some scores of additional groups undoubtedly exist. Much of this variety was imported from Europe. From the Reformation onward, Great Britain has accommodated many different Christian groups; and the massive immigration from the European continent contributed many others to the American population. In addition, several groups of Old World provenance underwent further evolution and splintering in the New World; and the United States in turn has produced, and continues to produce, a prodigious number of its own native Christian denominations, some of them, like the Mormons, Adventists, and Disciples of Christ, claiming large memberships. This multiplicity of religious bodies is visible not only at the national or regional scale but even within a single small city, village, or neighborhood. At the most obvious level of explanation, this may be a matter of the doctrine of separation of church and state or of an economy rich enough to support a superfluity of congregations and church buildings; but it is much more meaningful as a ringing affirmation of the dogged individualism of the American. The freedom to choose one's church, to float freely from one group to another, or even to invent an original sect of one's own is deeply cherished, and is closely akin to other forms of personal mobility and self-assertion.

[45] Church membership data are relied upon in this discussion not because they are ideal, or even satisfactory, indicators of the extent or intensity of religious practice but because no other measures are generally at hand. The strength of religious feeling on the part of individuals or groups is a phenomenon eluding easy quantification except for church attendance records, which are poor when they are available at all. Problems of American church membership and their interpretation are discussed in Wilbur Zelinsky, "An Approach to the Religious Geography of the United States: Patterns of Church Membership in 1952," *Annals of the Association of American Geographers*, 51, No. 2 (1961), 141–47. The discussion of denominational membership offered here is basically a brief paraphrase of this article. In addition, an indispensable fund of fact and interpretation is available in Edwin S. Gaustad, *Historical Atlas of Religion in America* (New York: Harper & Row, Publishers, 1962).

Yet, despite the luxuriation of creeds, it is also clear that religious feeling, at least in the conventional otherworldly sense, tends to run shallow in the United States. The incidence of church membership may be high and the financial investment in church property enormous, but traditional theological concerns scarcely ever enter the average citizen's head. At least, that has been the situation in recent decades, and the current is running strongly toward the further desanctification of thought and behavior. Such sacred occasions as Christmas, Easter, Passover, and the Sabbath have become largely secular and commercial, and church buildings have come to function more as community social centers than as houses of worship. Perhaps the most persuasive evidence for the casual role of the spiritual life in the American scheme of things is its relative invisibility in the physical landscape. In sharp contrast to the Christian, Moslem, or Hindu houses of worship of the Old World, or even those of neighboring Quebec and Mexico, the church building and associated activities are almost never located at the functional center of an American city, where land values are at their peak, but are relegated to peripheral sites.[46] Nor is tangible evidence of religion conspicuous in the country-side. And unlike most pre-modern communities of the Old World where daily existence and religion are mixed inseparably, the religious factor does not enter into the layout of town, field, or house, or the shaping of economic activities.[47]

On the other hand, in a new, emergent, perhaps unprecedented way, religion has become a potent force in the life of the country. There are hints of a novel mode of religion in the way in which imported creeds, including Roman Catholicism and Judaism along with the various brands of Protestantism, have become increasingly "Americanized," so that they now resemble one another much more than they do the same religions in the eastern Hemisphere. Furthermore, many of the more fundamentalist of the native churches, in what is not lightly called "God's Country," most vehemently identify themselves with extreme versions of patriotism.

Thus it can be said, without serious exaggeration, that Americanism has, in a most fundamental sense, become the true religion of Americans. The cult, which goes far beyond normal patriotism, is equipped with a full panoply of sacred writings, saints, rituals, and holy places. (The rise of Marxism-Leninism in the Soviet Union or Maoism in China offers some instructive parallels.) It would be difficult to find another object in human history so fiercely venerated as the American flag; and the heart-felt hysteria evident in flag-worship is a most convincing sort of religious testimony.

If Americanism does not fit neatly into the familiar mold of religion,

---

[46] The fact that Salt Lake City and Boston seem to be the only exceptions among larger American cities fortifies this contention, for these have been the metro-politan foci for the only two culture regions that begin to simulate theocracies.

[47] A theme pursued in detail in Pierre Deffontaines, *Géographie et Religions* (Paris: Gallimard, 1948) and David E. Sopher, *Geography of Religions* (Englewood Cliffs, N.J.: Prentice-Hall, Inc., 1967).

it may be that religion, like many other categories of human existence, is passing through a revolutionary set of changes as advanced societies enter a post-industrial age. The powerful strain of messianic perfectionism in the national character already alluded to is a major characteristic of this new faith (and of Marxism as well?). It is worth noting that the unabating vigor of North American missionaries, both Catholic and Protestant, in carrying the cross to Latin America, Africa,[48] Asia, and elsewhere is matched by their determination to spread the blessings of "the American way of life."

If virtually all the inhabitants of the United States are, consciously or not, adherents of a single religion in thought and deed, are the ordinary denominational classifications of any value to the cultural geographer? Most emphatically. When Americans categorize groups of people, the most important sets, beyond sex and age, are those taking into account race, religion, and ethnic group. If the racial boundaries are the most nearly impervious, the major religious faiths (Protestant, Catholic, Eastern Orthodox, and Jewish), are the next most stable, for it is more awkward to move or marry across interfaith lines than across almost any inter-ethnic line. Of central importance is the fact that religious associations seek out, accentuate, and preserve differences among persons. Not only can one's church affiliation denote national, ethnic, or racial origin, it also reflects rather strong urban-rural differentials, and in many parts of the nation the pervasive effects of regional culture. Thus there are significant correlations with larger spatial and temporal patterns. For that large minority of persons who elect to shift allegiance, the denomination may be a sensitive gauge of social, economic standing or aspiration, and psychological tendency.

There is some difficulty in providing usable maps of the major American religious denominations or derivative maps of religious regions (Figure 3.3) because of severe problems in defining membership and in procuring and evaluating statistics. But the results, imperfect though they be, are nonetheless instructive. It appears that, even though most denominational groups do tend, to a striking degree, to be national in distribution, no two groups display strongly congruent patterns, and each does have one or more major regional concentrations. These in turn can be attributed to specific events in cultural and demographic history.

More specifically, the spatial distribution of denominational members is the outcome of the following processes: births, deaths, migration, conversion (through casual contact or active missions), and apostasy. Although information is scanty, it seems that differential fertility and mortality are of relatively minor consequence in accounting for the size or distribution of religious groups. It is their spatial movements and success in attracting new members and holding old ones that are most critical. The contemporary map of American religion indicates four major sets of migrational events:

---

[48] Hildegard Binder Johnson, "The Location of Christian Missions in Africa," *Geographical Review*, 57, No. 2 (1967), 168–202.

Legend:

I — New England

II — Midland
II-a — Pennsylvania German

III — Upper Middle Western

IV — Southern
IV-a — Carolina Piedmont
IV-b — Peninsular Florida
IV-c — French Catholic
IV-d — Texas German

V — Spanish Catholic

VI — Mormon

VII — Western

Principal Catholic concentrations

FIG. 3.3. Major religious regions. (Adapted from: Wilbur Zelinsky, "An Approach to the Religious Geography of the United States," Annals of the Association of American Geographers, 51, No. 2 [June 1961], 193.)

1. the colonization of the Atlantic Seaboard by regionally distinctive groups;
2. their westward progression into the interior;[49]
3. the concentration of post-colonial immigrants from Europe, Canada, and Latin America in particular localities;
4. the rather less clearly visible interregional migration of members of different groups.

If the colonial settlers of British North America were an unrepresentative sample of Old World religionists, there was further discrimination in their regional destinations. The New England culture hearth was eminently receptive to Congregationalists (and later harbored Unitarians); the Midland was notably diverse in its religious makeup, but Quakers, Methodists, Presbyterians, Dutch Reformed, Lutherans, and various groups of Germanic origin were especially well represented, though not exclusive to the region. The early South was hospitable to all the major Protestant denominations of the period, except the more radical. Then, as migrants from these three nodal zones pushed westward, they also carried their religious identity with them, but with increasingly dilute regional distinctiveness as distance from native area increased. The nineteenth- and early twentieth-century immigrations can be detected wherever large clusters of Roman Catholics are counted, denoting as they do the Irish, various ethnic groups from eastern and southern Europe, Latin America, and Quebec. The presence of Scandinavians and German immigrants in many tracts of the Northern and Central United States is revealed by a high incidence of Lutherans; other Germans are traceable by the mapping of adherents of the Brethren, Evangelical and Reformed, and Evangelical United Brethren churches. Location of many immigrant church members over the past century and a half was governed by proximity to source or availability of suitable parcels of farmland. In one instance, that of the Jewish immigrant, however, there was a filtering downward through the hierarchical system of commercial metropolises, with the strongest representation toward the top of the pyramid in the larger cities.

The most intriguing questions in the spatial patterning of church members are suggested by differential rates of internal migration, conversion, and apostasy. If we have little hard data on migration differentials there is even less on differential ability to recruit and hold fresh converts. To put it in other terms, we know little about how effectively and rapidly the ideas in question diffuse through physical and social space. Gravitation toward certain points in the educational hierarchy is suggested by the strong concentration of Unitarians in all sizeable college communities, in addition to most large cities. There is also much evidence that frontier populations displayed particular affinities for the Presbyterian, Methodist, and Baptist creeds (the latter two very nearly American in origin); and there appears to be a peculiar compatibility

---

[49] As exemplified in Robert D. Mitchell, "The Presbyterian Church As an Indicator of Westward Expansion in 18th Century America," *Professional Geographer*, 18 (Sept. 1966), 293–99.

between Southerners and Baptism, especially after 1860. But what are we to make of the irregular map patterns displayed by Adventists, Disciples of Christ, or members of the Assembly of God and the Church of the Nazarene? The probability is that quite apart from accidents of ecclesiastical history, cultural forces we cannot yet identify have been at work in sorting out potential church members in terms of social and economic characteristics and preferences for particular areas and environments. Unfortunately, very little work has been done on the spatial array of religions at the micro scale in the United States, either within cities or in rural districts.[50] It is likely that even more puzzling questions of explanation and interpretation of the interrelations among location, religion, and other social factors would emerge from such documentation.

Further problems of theoretical interest are raised when regional boundaries are attempted on the basis of religious affiliation (Figure 3.3, p. 97). A total of seven primary regions are proposed, plus an additional five subregions, in an essay by this writer.[51] How closely do these regions resemble the more general culture areas delimited in the next chapter? Rather well in at least four instances. The Mormon realm is an admirable example of a region in which religion is the chief genetic factor as well as the strongest reason for a persistent distinctiveness. The special regional character of New England might also be attributed in part to its quasi-theocratic past and the unique blend of denominations housed within its boundaries. And religion is certainly a significant element in the maintenance of cultural identity within the Hispanic Southwest. In religion, as in many other respects, the Southern culture area tends to stand apart from the rest of the nation, although this distinctiveness seems to have materialized rather recently. This is because of the relatively high incidence of certain Protestant denominations, especially the Baptist, and the scarcity of Catholics and Jews, rather than because of any uniquely Southern group. On a lesser scale, an argument can be advanced for the Pennsylvania German area having acquired its unique geographic personality mainly because of its large quota of "peculiar people" and other staunch church members for whom religion has been a dominant force. On the other hand, the Middle West fails completely to assert itself in terms of a distinctive constellation of religions, and so too the West and its subregional components, excepting, of course, the Mormon area.

Thus, although we can do no more than lightly sketch the possibilities here, the spatial array of denominational members in the United States would seem to hold considerable potential in illustrating one major set of processes at work in shaping the more general cultural map: differ-

---

50 One of the rare specimens of this genre is Elaine M. Bjorklund, "Ideology and Culture Exemplified in Southwestern Michigan," *Annals of the Association of American Geographers*, 54, No. 2 (1964), 227–41.

51 Zelinsky, "An Approach to the Religious Geography of the United States," Fig. 26, p. 193. This initial attempt at regionalization should be used in conjunction with a more sophisticated drawing, Fig. 8, "Religious Affiliation in the United States, ca. 1950," Sopher, *Geography of Religions*, pp. 84–85. This contains additional information and a quite different mode of symbolization.

ential migration, or diffusion, of individuals and ideas at different periods. At a somewhat deeper level, it may offer raw material for the understanding of the still mysterious processes of region formation. And, finally, if we dare venture far beneath the arithmetical surface of denominations, members, and church buildings into the inner workings of religion in the larger sense, we may be witnessing one facet of a major structural transformation as a new mode of religious sensibility seizes the subconscious of a folk.

Other Aspects of American Culture

If the analysis of American house types and religious patterns can yield major insights into the nature and spatial behavior of the national culture, there are many other cultural items that might be eminently serviceable for the same purposes. Each reveals a special perspective, and none can be ignored without dimming our view of the vast, convoluted landscape that is the American cultural system. The following paragraphs include a listing of the more immediately profitable topics, with brief indications of what has already been done and, rather more pointedly, what can or should be done. Lack of space, however, precludes full-scale critiques or programmatic statements. In some sectors, practically no serious work has been executed by geographers or anyone else; but even where a significant body of work is at hand, as in the study of folk housing or religion, the amount of unfinished business still bulks immensely larger.

SETTLEMENT FEATURES. The settlement landscape offers a peculiarly attractive array of phenomena for geographic analysis, for it combines the effects of culture and tradition with those of habitat, economy, and technology in highly mappable ways. In addition to farmhouses, which until the turn of this century accommodated the great majority of the American population, there are the many varieties of barns, sheds, wells, and other farmstead structures meriting close attention. Enough work has been done on barns to indicate that they are perhaps the most conservative of structures in general use and may serve quite as effectively as house types to indicate sources, routes, and dates of settlers and the spatial outlines of culture areas.[52] Distinctive barn types are common to the South, others to the Midland and French Canada. Much of older New England is characterized by the readily recognizable connecting barn.[53] Enough work has been done on the fencing (wooden,

---

52 The topic of barns as indices to early cultural identity is treated in Henry Glassie, *Pattern in the Material Folk Culture of the Eastern United States* (Philadelphia: University of Pennsylvania Press, 1968), pp. 60–64, 88–92, 133–41, 145–49; Eric Sloane, *American Barns and Covered Bridges* (New York: Wilfred Funk, Inc., 1964), and Joseph W. Glass, *The Pennsylvania Culture Region: A Geographic Interpretation of Barns and Farmhouses*, doctoral dissertation, Pennsylvania State University, 1971.

53 A comprehensive bibliography for American barns is scattered through Glassie, *Pattern in Material Folk Culture*. For French Canada, see Robert-Lionel Séguin, *Les Granges du Québec du XVIIe au XIXe Siècle*, Bulletin No. 192 (Ottawa:

stone, stump, sod, hedging) used to bound fields and pastures to reveal some sharp regional differences and the probability that further research would be profitable, but not enough to suggest the nature of the ultimate findings.[54] A single study of the historical geography of the covered bridge is an admirable example of the leadership role of New England in devising new technologies, and of the mechanics of diffusion as well.[55] It also suggests the value of analyses of other bridge types or of the grist mills and country stores that punctuate the rural scene at frequent intervals. Another universal element in the American settlement landscape, rural and urban, the church building, has been scandalously neglected by geographers. The architectural historians have confined themselves to the grander, professionally designed specimens, leaving vernacular church design in limbo. As an item responsive to ethnic, racial, temporal, socioeconomic, and urban-rural, as well as denominational, differentials, the church building would seem to represent too rich an ore to be overlooked by cultural geographers or historians. There has been an equally puzzling avoidance of cemeteries by students of Americana, despite the abundance of field and cartographic material and the anthropological axiom that burial customs are among the more revealing departments of human behavior. The entire literature has been noted in a very brief statement that is also a forceful manifesto for the research that is so badly needed.[56]

In settlement studies, as elsewhere in American cultural geography, there has been a strong bias toward the rural and agricultural, so that scant attention has been paid to the urban scene and its artifacts. Thus we have virtually nothing on the historical geography or cultural implications of factory buildings, warehouses, or shops, or of the public buildings and spaces usually situated in towns (for example, schools, jails, depots, post offices, hospitals, hotels, zoos, theaters, parks, and other amusement places), or of the statues or other monumental constructions of such rich symbolic import. And such work as has been done on house types has largely ignored the specifically urban forms.[57]

The spatial configuration of settlement features has received at least modest attention from geographers, enough at least to indicate the promise of ampler rewards. The historical geography of land survey has

---

Musée National du Canada, 1963); for New England, Wilbur Zelinsky, "The New England Connecting Barn," *Geographical Review*, 48, No. 4 (1958), 540–53; and for the Midland, Alfred L. Shoemaker, ed., *The Pennsylvania Barn*, 2d ed. (Kutztown, Pa.; Pennsylvania Folklife Society, 1959); and Glass, *The Pennsylvania Culture Region.*

[54] J. Fraser Hart and E. C. Mather, "The American Fence," *Landscape*, 6, No. 3 (Spring 1957), 4–9; Wilbur Zelinsky, "Walls and Fences," *Landscape*, 8, No. 3 (Spring 1958), 14–20; and Glassie, *Pattern in Material Folk Culture*, *passim*.

[55] Fred Kniffen, "The American Covered Bridge," *Geographical Review*, 41, No. 1 (1951), 114–23.

[56] Fred Kniffen, "Necrogeography in the United States," *Geographical Review*, 57, No. 3 (1967), 426–27.

[57] The situation is thoroughly reviewed in Kenneth E. Corey, *A Spatial Analysis of Urban Houses*, doctoral dissertation, University of Cincinnati, 1969.

been usefully sketched by Marschner,[58] and the establishment of rectangular survey and a related road system in Ohio has been probed in some depth.[59] Only a single serious effort seems to have been made to examine the geography of field boundaries,[60] in contrast to the popularity of the topic among European scholars; but additional work on field, road, and survey patterns would probably repay the investigator many times over. Similarly, with the analysis of farmstead patterns, on which topic the pioneering essay still stands in splendid isolation.[61] Price's innovative study of the pattern of American county courthouse squares and their distribution through time and space has documented a most interesting series of diffusional processes and their impact upon urban morphology.[62] But, again in odd contrast to the European record, there are few other attempts to scrutinize urban street patterns. Such studies as we have indicate that the geometry of street layout (and the patterning of street names) may serve as effective tracers of cultural flows and one of the more reliable means for delimiting major culture areas.[63]

Casting a rather broader observational net are those geographers and other scholars who have looked at the composite visible landscape in the United States and tried to interpret it. In some local studies, such as those of Nelson on Southern California[64] or Swain and Mather on the "St. Croix Border Country,"[65] the cultural element is largely implicit. A number of recent critiques of the appearance of contemporary America, often sad or scornful in tone, have summoned up visual evidence in indicting some of the central values of the culture.[66] But other observers,

[58] Francis J. Marschner, *Land Use and Its Patterns in the United States,* Agriculture Handbook No. 153 (Washington, D.C.: United States Department of Agriculture, 1959).

[59] Norman J. W. Thrower, *Original Survey and Land Subdivision: A Comparative Study of the Form and Effect of Contrasting Cadastral Surveys* (Chicago: Rand McNally & Company, 1966); and William D. Pattison, *Beginnings of the American Rectangular Land Survey System 1784–1800,* Research Paper No. 50 (Chicago: University of Chicago, Department of Geography, 1957).

[60] J. Fraser Hart, "Field Patterns in Indiana," *Geographical Review,* 58, No. 3 (1968), 450–71.

[61] Glenn T. Trewartha, "Some Regional Characteristics of American Farmsteads," *Annals of the Association of American Geographers,* 38, No. 3 (1948), 169–225.

[62] Price, "The Central Courthouse Square in the American County Seat."

[63] John W. Reps, *The Making of Urban America: A History of City Planning in the United States* (Princeton, N.J.: Princeton University Press, 1965); and Richard Pillsbury, "The Urban Street Pattern As a Culture Indicator: Pennsylvania, 1682–1815," *Annals of the Association of American Geographers,* 60, No. 3 (1970), 428–46.

[64] Howard J. Nelson, "The Spread of an Artificial Landscape over Southern California," *Annals of the Association of American Geographers,* 49, No. 3, Part 2 (1959), 80–99.

[65] Harry Swain and Cotton Mather, *St. Croix Border Country* (Prescott, Wisc.: Trimbelle Press and Pierce County Geographical Society, 1968).

[66] Peter Blake, *God's Own Junkyard: The Planned Deterioration of America's Landscape* (New York: Holt, Rinehart & Winston, Inc., 1963); Ian Nairn, *The American Landscape: A Critical View* (New York: Random House, Inc., 1965); and Ian L. McHarg, *Design with Nature* (Garden City, N.Y.: Natural History Press, 1969).

such as Jackson, Kouwenhoven, and Tunnard and Pushkarev,[67] have been more temperate in their judgments; and Lowenthal and Prince's dictum that "Landscapes are formed by landscape tastes"[68] would seem a useful point of departure for the culturally attuned landscape-watcher, and for those concerned with the newly crystallized subfield of perceptual geography.

CULTURAL DEMOGRAPHY AND CULTURAL ECONOMICS. Demographers and geographers dealing with population matters have tended to restrict their explanations almost totally to traditional social and economic factors. But it would appear that in the United States, where fertility, mortality, and migrational differentials are not so largely shaped by economic considerations as in less affluent societies, more attention could be given to other modes of explanation. These could include not just the search for pleasure and the amenities, but also the role of cultural differences, regional and ethnic. Although such an approach might be especially pertinent in coping with migration, there are also interesting possibilities in terms of morbidity and causes of death.

To illustrate the point, consider a simple map showing homicide rates among the white population during a recent period. (See Figure 3.4.) The marked interstate differentials displayed in this drawing are not randomly distributed, but reveal instead a distinct variance by region. The risk of homicide is least in New England and the Northeast in general, and in those states that received major inputs of migrants and ideas from New England. The much higher rates in most of the South and West (with the interesting exception of Utah) may indicate a generally lower level of socioeconomic attainment but even more strongly the dilution of a New England cultural predisposition against personal violence. Unfortunately, earlier data to test this notion are not readily available.

Within the well-trodden field of economic geography, emphasis has consistently fallen on questions of production and distribution. Neither the statistical systems nor the curiosity of students has been directed toward the equally interesting problems of consumption. If, as we may legitimately speculate, cultural factors affect consumption in many ways, there has been a disturbing dearth of interest in patterns of food or beverage consumption, or in any aspect of dietary geography.[69] The few probes into this field are most provocative;[70] and again it seems

[67] Christopher Tunnard and Boris Pushkarev, *Man-Made America: Chaos or Control? An Inquiry into Selected Problems of Design in the Urbanized Landscape* (New Haven: Yale University Press, 1963); and John A. Kouwenhoven, *The Beer Can by the Highway: Essays on What's American about America* (Garden City; N.Y.: Doubleday & Co., Inc., 1961). John B. Jackson's observations are scattered through many numbers of *Landscape,* usually in the form of unsigned editorials.

[68] David Lowenthal and Hugh C. Prince, "English Landscape Tastes," *Geographical Review,* 55, No. 2 (1965), 186.

[69] Herbert G. Kariel, "A Proposed Classification of Diet," *Annals of the Association of American Geographers,* 56, No. 1 (1966), 68–79; and Max Sorre, "La Géographie de l'Alimentation," *Annales de Géographie,* 51 (1952), 184–99.

[70] Rupert B. Vance, *Human Geography of the South* (Chapel Hill: University

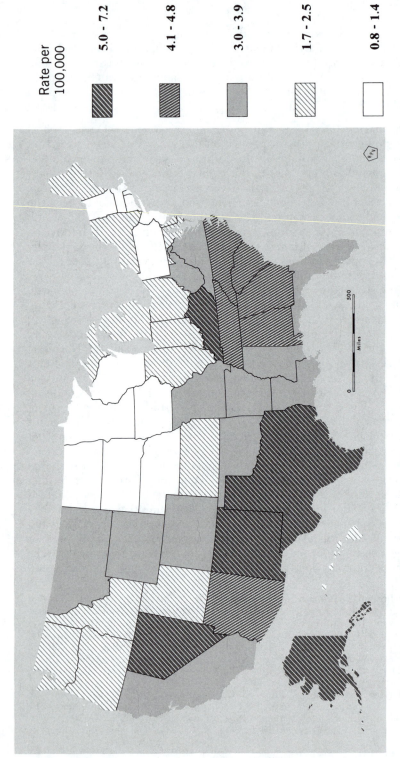

FIG. 3.4. Age-adjusted homicide rate, white population, three-year average, 1959–1961. (Adapted from: National Center for Health Statistics, Homicide in the United States, 1950–1964, Public Health Service Publication No. 1000, Series 20, No. 6 [Washington, D.C., 1967], Table 2, pp. 16–17.)

Rate per 100,000

5.0 - 7.2

4.1 - 4.8

3.0 - 3.9

1.7 - 2.5

0.8 - 1.4

clear that the intellectual payoff would be considerable for the geographic pioneer with the courage and persistence (and intestinal fortitude?) to assemble and analyze the disparate, often refractory data. The same statement could be repeated for other consumer items, such as clothing, cosmetics, furniture, jewelry, color preferences, and the like, where choices are dictated by emotion, tradition, apparent whim, or other "irrational" factors. The data presented in Table 3.1, subject though they are to large sampling errors and despite the grossness of the areas, suggest some rich possibilities. As in the field of electoral geography, it is almost certain that valuable data have been compiled and analyzed by advertising, manufacturing, wholesale, and retail corporations; but, if so, most of it remains inaccessible to the public.

LANGUAGE. The one aspect of American culture that has been approached diligently in organized fashion—and with due regard to the spatial dimension—is the vocabulary, pronunciation, and syntax of American English. This has been the accomplishment largely of a group of linguists, with only fitful help from geographers.[71] The resultant maps quite clearly indicate a close areal correspondence between the various American dialects and subdialects and other phases of culture, material and nonmaterial. Linguistic areas thus have proved to be excellent surrogates for general culture areas in the United States, as in so many other parts of the world. One can only hope that these linguistic endeavors will continue and reach the degree of sophistication and detail achieved in European research.

As might be expected, the chief contribution by geographers to this study area has been on the topic of place names.[72] The study of the

---

of North Carolina Press, 1935), 411–41; George R. Stewart, *American Ways of Life* (Garden City, N.Y.: Doubleday & Co., Inc., 1954), 75–132; and Sam Hilliard, "Hog Meat and Cornpone: Food Habits in the Ante-Bellum South," *Proceedings of the American Philosophical Society*, 113 (January 1969), 1–13. The potential contribution of dietary studies to our understanding of the regional cultures of the United States is illustrated in Don Yoder, "Historical Sources for American Traditional Cookery: Examples from the Pennsylvania German Culture," *Pennsylvania Folklife*, 20, No. 3 (Spring 1971), 16–29.

[71] Most of the research effort by linguists on region aspects of American speech has been devoted to an ongoing, massive project, *The Linguistic Atlas of the United States*. The status of this work and the publications issued thus far are described in Carroll E. Reed, *Dialects of American English* (Cleveland: The World Publishing Company, 1967). The more noteworthy studies already published include Hans Kurath, ed., *Linguistic Atlas of New England*, 3 vols. (Providence, R.I.: Brown University, 1939–1943); and Hans Kurath, *A Word Geography of the Eastern United States* (Ann Arbor: University of Michigan Press, 1949) and *The Pronunciation of English in the Atlantic States* (Ann Arbor: University of Michigan Press, 1962).

[72] For example, Robert C. West, "The Term 'Bayou' in the United States: A Study in the Geography of Place Names," *Annals of the Association of American Geographers*, 44, No. 1 (1954), 63–74; Wilbur Zelinsky, "Some Problems in the Distribution of Generic Terms in the Place-Names of the Northeastern United States," *Ibid.*, 45, No. 4 (1955), 319–49; and Meredith F. Burrill, "Toponymic Generics," *Names*, 4, No. 3 (September 1956), 129–45 and 4, No. 4 (December 1956), 226–40.

Table 3.1. Regionally Distinctive Patterns of Consumer Expenditures for Selected Goods and Services, 1960-1961

Based upon Sample of 12,000 Nonfarm Households
National mean household dollar expenditure per item = 100

	Northeast	North Central	South	West
Flour	60.4	75.4	166.1	85.7
Macaroni, spaghetti, noodles	160.4	89.4	61.0	97.2
Rice	100.9	54.7	108.9	151.5
Cornmeal	22.7	45.3	259.3	30.1
Bread other than white	155.9	85.0	44.1	165.7
Beefsteaks, fresh and frozen	132.1	84.9	80.1	105.9
Bacon	79.2	94.3	120.3	105.9
Cold cuts	133.8	111.0	67.7	82.8
Smoked sausage	96.2	118.8	103.3	76.4
Fresh, frozen, smoked fish and seafood	162.2	66.0	76.3	97.1
Canned tuna	141.5	73.4	66.1	131.4
Fresh whole milk, retail	145.2	103.7	59.2	94.2
Pineapple juice	150.9	62.2	77.9	117.1
Fresh snap beans	107.5	58.4	164.4	42.8
Frozen green beans	183.0	60.3	50.8	117.1
Frozen lima beans	107.5	60.3	142.5	77.1
Frozen spinach	162.2	69.8	79.6	85.7
Butter	149.0	126.4	37.2	91.4
Margarine	92.4	90.5	106.7	114.2
Lard	43.3	66.0	210.1	51.4
Sugar	92.5	94.1	120.0	85.9
Syrup, molasses, honey	92.5	85.0	115.0	104.1
Tea	143.6	67.9	105.0	74.0
Cola drinks	75.4	92.4	145.9	71.1
Gingerale	220.8	73.8	35.4	65.6
Olives	136.0	105.9	50.8	120.0
Salt, other seasonings	94.2	83.0	105.0	116.9
Beer and ale	141.8	105.9	55.9	102.2
Blended whisky	173.7	100.0	50.8	71.1
Bourbon, scotch, straight rye	105.1	75.8	76.0	145.9
Cigarettes	114.8	96.2	91.5	97.0
Cigars	145.8	87.6	79.7	82.8
Other tobacco	77.5	73.8	174.2	48.6
Moving, freight, and storage	79.5	85.1	93.0	167.8
Writing supplies, stationery	114.7	98.0	75.9	120.0
Insect sprays, powders	92.0	53.8	133.1	125.9
Air fresheners, deodorizers	105.3	72.4	94.9	104.1
Electric blankets	60.1	61.1	114.0	197.0
Hats and caps, (men 18 years and over)	111.9	102.0	105.1	71.4
Slacks, shorts, dungarees (girls 16 and 17)	140.0	100.0	81.1	71.4
Hats (girls 16 and 17)	162.0	100.0	83.0	34.2
Gloves (girls 16 and 17)	128.2	114.8	94.6	60.0
Motorcycles and scooters	47.1	92.6	101.8	182.8
Indoor movies	130.1	87.0	74.1	108.5
Drive-in movies	83.0	89.0	101.8	140.1
Sports events	86.9	103.9	91.4	123.1
Concerts, plays, other admissions	170.1	85.2	47.2	108.5
Crafts, other hobbies	115.2	98.4	53.2	157.2
Pet food	117.1	78.0	87.9	128.7
Newspapers	132.1	103.9	74.1	102.9
Pocket edition books	109.4	81.6	79.1	134.1
Comic books	103.9	81.6	101.8	117.8
Hardbound books	89.0	89.0	96.3	130.1

Adapted from Fabian Linden, ed., *Market Profiles of Consumer Products* (New York: National Industrial Conference Board, 1967).

usage of generic terms singly or in groups over considerable tracts has proved especially fruitful in revealing the interaction of man and habitat and the spatial flow of ideas. There has been less interest in the geography of specific names; but an analysis of one particular type, classical town names, has documented nationwide, but irregular, dispersal from a west central New York hearth of a trait that is particularly diagnostic of the national culture.[73] Clearly there is much more to be learned by exploring our place name cover, whether through intensive local study[74] or through study of single terms, or classes thereof, over wider areas.

INSTITUTIONS, SOCIAL BEHAVIOR, AND FOLKLORE. The geography of social institutions is, potentially at least, a subject of considerable interest to the student of American culture. Unfortunately, there has been virtually no work done to date on the origin, spread, and geographical significance of such items as kinship groups; fraternal organizations; colleges; scientific, trade, and professional associations; and any of a vast array of other voluntary associations. A single attempt in this direction, an essay on the spatial career of county agricultural fairs, serves to whet the appetite.[75]

The record is equally blank as to social behavior, a field in which the interests of geographer, folklorist, sociologist, and psychologist intersect. Political geography, and more specifically electoral geography, might be included here, since the cultural overtones are loud; but it has begun to develop so vigorously of late that a brief note could not do it justice.[76] Among many other items, it can be suggested that the cultural geographer would be well rewarded by attending to all the many forms of recreation—field sports, children's games, card games, and holiday celebrations, among others. And, beyond that, there is the vast, enticing, and motley field known as folklore. It will suffice for the moment simply to list some of the many fertile topics that bear looking into: folk tales and legends; proverbs; riddles; superstitions; folk music; folk dance; handicrafts; gestures; social customs; and the many varieties of farm, household, and craft utensils.[77]

[73] Wilbur Zelinsky, "Classical Town Names in the United States: The Historical Geography of an American Idea," *Geographical Review*, 57, No. 4 (1967), 463–95.

[74] For a model study of this type, see Frederic G. Cassidy, *The Place-Names of Dane County, Wisconsin*, Publications of the American Dialect Society, No. 7 (Greensboro, N.C.: American Dialect Society, 1947).

[75] Fred Kniffen, "The American Agricultural Fair: Time and Place," *Annals of the Association of American Geographers*, 41, No. 1 (1951), 42–57.

[76] A good overview of the situation is afforded in the recent exchange of views between Cox and Kasperson: Kevin R. Cox, "Suburbia and Voting Behavior in the London Metropolitan Area," *Annals of the Association of American Geographers*, 58, No. 1 (1968), 111–27; Roger E. Kasperson, "On Suburbia and Voting Behavior," *Ibid.*, 59, No. 2 (1969), 405–11, and Kevin R. Cox, "Comments in Reply to Kasperson and Taylor," *Ibid.*, 59, No. 2 (1969), 411–15. For an ambitious survey—of sorts—of the political landscape of the nation, with heavy emphasis on ethnic and regional factors, see Kevin P. Phillips, *The Emerging Republican Majority* (Garden City: Doubleday & Co., Inc., 1969).

[77] Excellent introductions to the field are available in Richard M. Dorson, *American Folklore* (Chicago: University of Chicago Press, 1959) and Jan Harold Brunvand, *The Study of American Folklore: An Introduction* (New York: W. W.

The preceding pages may leave the reader with the impression that cultural geographers are enamored with quaint and arcane phenomena, and more excited over what is old, local, or special than in the pursuit of grander scientific quarry. This is at most a half-truth. In sampling specific topics that offer major potential for illuminating basic processes of cultural change through space and time, we have been obliged, for the sake of convenience, to divide what is really a whole. When the deeper layers of reality are probed, we can see that place names, barns, fences, diet, phonetics, voting patterns, burial customs, churchgoing, and much else are all related, interlocking phenomena, deeply imbedded within the immense matrix of culture. These seemingly disparate things can be the means to an end rather than the end of cultural research. In lieu of any options for experimenting in the laboratory or for applying strict deductive logic to the still obscure, but plainly very complex, processes at work in cultural change, the inductive route seems the most efficient strategy for identifying general cultural laws. And even though the job of systematically gathering, sifting, and collating information on a variety of spatio-temporal phases of American culture has barely begun, some hints of general process and larger structure are beginning to appear. For the geographer intent upon linkages among outwardly different items, and the larger designs of human thought and action through space and time, there is something particularly intriguing about the tendency for cultural patterns to fall into reasonably uniform territorial packages. In the final chapter, we examine such culture areas as they exist in the United States, and begin to explore some of the potentialities for expanding our theoretical frontiers in following such an inductive approach.

Norton & Co., Inc., 1968). One of the rare explicitly geographic ventures into American folklore is E. Joan Wilson Miller, "The Ozark Culture Region As Revealed by Traditional Materials," *Annals of the Association of American Geographers*, 58, No. 1 (1968), 51–77.

CHAPTER FOUR

Structure

Why Study American Culture Areas?

The cultural processes that have operated so vigorously in the Europeanized portions of North America over the past three and a half centuries have given us a set of reasonably homogeneous, contiguous tracts of territory whose inhabitants are at least dimly aware of a common cultural heritage and of differences from other territorial groups. Such tracts we can properly label "culture areas," even though they differ in some notable ways from the standard, classical culture areas of long-settled, slowly evolving communities elsewhere in the world. This chapter is based on the premise that the existence of these regions is a large, truly significant fact in the human geography of this nation, and that their analysis is a rewarding activity for the scholar in both a theoretical and a practical sense. At the most naïve level, Americans are spontaneously curious about the local peculiarities of their compatriots. Thus a favorite theme for social small talk is the distinguishing characteristics of the Yankee, Southerner, Hoosier, New Yorker, Texan, Pennsylvania German, or Southern Californian. At a more sophisticated level, it is plausible that the character of these more or less coherent spatial groupings of people has had a palpable effect upon social, economic, and political behavior at both the local and national levels, and that we need to know more about these relationships. An ultimate objective is to gain insight into some basic problems of culture theory.

In general, the differences among American culture areas tend to be slight and shallow as compared to most older, more stable countries, even though in one instance, that of the South, they helped precipitate the gravest political crisis and bloodiest military conflict in the nation's history. The often subtle nature of interregional cultural differences can be ascribed to the recency of American settlement, a perpetually high mobility rate, an elaborate communications system, and centralization of economy and government. Indeed a case might be built against the study of American culture areas on the grounds that they are quaint vestiges of a vanishing past, soon to be expunged from the living landscape and of interest only to antiquarians. But even if one concedes the likelihood that powerful forces are pushing many—but far from all!—departments of American thought and behavior toward standardization, the lingering effects of the older culture areas are still very much with us; and there is strong evidence that newer, quite potent ones are arising. Thus, for example, more than a century after the Civil War, the South remains a powerful political and social entity, and its peculiar status is formally recognized in religious, educational, athletic, and literary circles. In some instances indeed, the gap between South and non-South may be widening. But even more intriguing is the appearance of a series of essentially twentieth-century regions. The Southern Californian is the largest and most distinctive, and its quite special regional culture has yet to reach full bloom. Similar trends are visible in southern Florida and in the burgeoning Texas identity, and probably in the more flourishing segments of New Mexico and Arizona as well. At the metropolitan level, it is difficult to believe that such aggressively narcissistic places as San Francisco, Las Vegas, Dallas, Tucson, or Seattle are becoming that much like all other American cities. At the micro scale, some less obtrusively embryonic areas will be cited later. If there is any specific message in this chapter, it is that the subdued, yet meaningful, internal cultural heterogeneity of an older America is being supplanted by a novel mosaic, equally variegated but pieced together from newer materials and with new forces.

The Classification and Anatomy of Culture Areas

The culture areas of the world, past, present, and future, can be usefully divided into two classes on the basis of mode of origin. The first, and by far the more numerous, since it accounted for virtually all culture areas until the onset of the modern period and the emergence of some truly advanced, radically transformed populations, is perhaps best termed the *traditional region*. These regions are relatively self-contained, endogamous, stable, and of long duration. The individual is born into the region and remains with it, physically and mentally, since there is little in- or out-migration by isolated persons or families; and the accidents of birth would automatically assign a person to a specific caste, class, occupation, and social role. An intimate symbiotic relationship between man and land develops over many centuries, one that creates indigenous modes of thought and action, a distinctive visible landscape, and a form of human ecology specific to the locality. Although the usual processes of random cultural mutation, the vagaries of history, some slight intermixture of peoples, and the diffusion of innovations of all sorts prevents the achievement of total stasis or equilibrium, or complete internal uniformity, it would not be unfair to characterize such a traditional region as one based upon blood and soil. In the extreme, it becomes synonymous with a particular tribe or ethnic group.

The second class of culture area, the *voluntary region*, is still emergent, hence not yet clearly recognized. It is coeval with the process of modernization in Northwest Europe and the appearance of individuals who are essentially free agents, in spatial and other dimensions. It is in North America that the earliest, largest, and, to date, most advanced experimentation in devising voluntary regions has gone on. Our thesis is that the traditional spatial and social allocation of individuals through the lottery of birth is being replaced gradually by a process of relative self-selection of life style, goals, social niche, and place of residence.

This new, indeed revolutionary, mechanism can be detected from the earliest days of serious North American settlement onward. The colonists were motivated by a variety of considerations—religious, social, political, economic, and idiosyncratic. With some notable exceptions, however, they traveled as individuals or single families, gravitating toward those places perceived as best fulfilling their aspirations, and where they could hobnob with many strangers, of widely scattered origin but with similar tastes and proclivities. The entire process is beautifully exemplified in the founding of the Great Basin Kingdom by the Mormons. There we see a thoroughly self-conscious (and amply documented) routing into a remote, empty tract of like-minded persons of highly diverse background bent upon building a tightly structured, homogenous society. The same trends were at work in the occupation of the initial colonial nodes along the Atlantic Seaboard and later in the advancement of the frontier, but rather less visibly and distinctly, in part because of a large amount of random migrational jostling.

Once an American area had been effectively settled and had achieved some semblance of economic and social stability, there began a series of processes analogous to, but less intense than, those typical of the traditional region. Emotional and social roots were sunk into the land; man and habitat interacted in endlessly complex fashion; diverse people met and merged their ideas and ways of life; influences were given to and received from the outside. In short, a new culture area had crystallized. The older American case simulates, but cannot match, the premodern culture area in purity and longevity. With adequate isolation or richness of development, the regional culture might develop a persistent vitality of its own. But frequently the same forces of individualistic choice that led to its birth also promote its erosion and obliteration. The more enterprising elements in the population drift away to greener pastures; and the relict culture area may become the stage upon which new forces operate and newer voluntary regions take shape. It is altogether possible today to be a solid citizen of metropolitan New England or Philadelphia and yet be utterly oblivious to the historic cultural role of those places.

In late twentieth century America, the currents of voluntary migration are many and varied; they remain preeminently economic in character, but increasingly the quest for pleasure and the amenities plays a role in the decision to migrate. One might argue forcefully that the spatial sum of millions of such decisions in a mass society is toward an un-relieved sameness from coast to coast, the progressive blotting out of local deviation, especially in urban areas.[1] But a growing mass of evidence points in just the opposite direction: the spawning of new families of voluntary regions, quite distinct in shape and structure from the older traditional regions. Some of these regions are ephemeral in the extreme, others have the promise of durability, but, in the aggregate, they amount to a new geometry of culture, as mobile individuals seek ideal havens on their journey of self-discovery.

The argument underlying this chapter, then, is that the system of culture areas observable in contemporary America is dual in character. There is the older set of places, originating during and just after the period of First Effective Settlement. Some of these are relict in character, others still prospering, but they are all hybrid creatures in the way they combine elements of the traditional areas with those of the voluntary. Partially superimposed over the older set are the many new voluntary regions, some still quite indistinct, usually with scarcely any functional connection with their predecessors.

A third, and trivial, type of region might be noted, then summarily dismissed: the spurious or synthetic region. With two or three marginal exceptions, we can so classify all 48 of the conterminous United States, despite the strong popular inclination to think of each as somehow a distinctive package of people. Falling into the same company are the concoctions of the advertising agency and other vested interests—"Chicagoland," Delmarva, TVA Country, "Pittsburgh Pirate Land," and the like. Rather more interesting are the attempts of commercial promoters to breathe new life into genuine, but fading, cultural entities: the Potemkin Village-like re-creations along superhighways of what the passing tourist might expect Pennsylvania German country to look like, or the largely fictitious simulations of frontier cowboy culture in towns that mock reality by imitating the never-never land of movies and television.

Let us define a culture area as a naïvely perceived segment of the time-space continuum distinguished from others on the basis of genuine differences in cultural systems.[2] The two characteristics that set it aside from other varieties of geographic region are: (1) the extraordinary number of ways in which it is manifested physically and behaviorally; and (2) the condition of self-awareness on the part of participants. The former trait has already been discussed, and the point made that no other discrete category of geographic region, short of that elusive entity, the holistic region, contains such a great variety of phenomena. The latter characteristic is unique. One must insist that if self-consciousness is

---

[1] An idea advanced vigorously in Melvin M. Webber, "Culture, Territoriality, and the Elastic Mile," *Regional Science Association, Papers,* 13 (1964), 59–69, and in Alvin Toffler, *Future Shock* (New York: Random House, Inc., 1970), pp. 91–94.

[2] Joseph E. Schwartzberg, "Prolegomena to the Study of South Asian Regions and Regionalism," in Robert I. Crane, ed., *Regions and Regionalism in South Asian Studies: An Exploratory Study,* Comparative Studies on Southern Asia, Monograph No. 5 (Durham, N.C.: Duke University Press, 1967), p. 93.

lacking (and it will suffice if it is present only at the subliminal level), then we are examining something other than a genuine culture area. The recognition of the area and one's affiliation with it can run the gamut from a raging sense of nationhood to fleeting, unemotional perceptions of small gradations of differences between one's local way of life and another.

A multi-tiered, nested hierarchy of culture areas might be postulated for the United States; but the most interesting levels, and those for which we can adduce the greatest number of general geographic principles, are the two topmost ones: the nation as a whole, and the five to ten (I prefer the former) of subnational regions, each comprising parts or all of several states. Given the fact that Americans have been vigorously expansionist all during their history, it is not surprising to find that the citizenry has filled up virtually the whole of the national space and spilled out far beyond into many military, diplomatic, missionary, scientific, and touristic exclaves. But, excluding such interesting outliers, which are scattered over much of the globe, and also various enclaves of unassimilated aboriginal or immigrant groups, there is a remarkably close coincidence between the conterminous United States and the cultural United States. In fact, our international boundaries present the neatest cultural cleavages to be found within North America. As the traveler crosses from this country into Mexico, he is negotiating a visible passage across a veritable cultural chasm, despite the intense Americanization of northern Mexico and the quite apparent Hispanicization of our Southwest border zones. If the contrasts are less dramatic between the two sides of the United States-Canadian boundary, they are nonetheless real, especially to the Canadian. Our northern border separates not only two distinct, if interrelated, political and economic systems, but also two different, yet closely fraternal, modes of cultural behavior. There is the abrupt, immediate change, for example, from one system of English pronunciation to another, and from one style of folk architecture to something noticeably alien. Undoubtedly other discontinuities could be recorded in other departments of culture. There is a major erosion of the cultural barrier along only a single major segment of our international frontier: that reaching from northern New York to Aroostook County, Maine. There a vigorous demographic and cultural invasion by French Canadians from Quebec and New Brunswick has gone far toward eradicating previous international differences.

But if the international boundaries are potent and precise as a cultural container, the interstate boundaries, on the other hand, are curiously irrelevant. In large part, this results from the fact that most were the relatively arbitrary creations of the federal Congress. But even when the state enjoyed a strong autonomous early existence, as did Massachusetts, Virginia, or Pennsylvania, subsequent economic and political forces have been powerful solvents, softening and washing away initial identities. In fact, our present set of 48 conterminous states is so anachronistic in the context of contemporary socioeconomic realities, not to mention cultural forces, that they could be readily scrapped and replaced by another, much better set were it not for the deterrent effect of inertia and various vested interests. And having served so long as spatial pegs (like professional baseball or football teams) on which we can hang signs stating our identity, however speciously, or as ersatz outlets for whatever vestigial "territorial instinct" lingers on in us, the wiping out of our state boundaries would set off a mighty uproar. Halfway convincing cases might be built for equating the states of Utah and Texas with their respective culture areas because of exceptional historical and physical circumstances—or perhaps Oklahoma, given a tardy European occupation and unique status as dumping ground for all the relict Indian groups of the East. But in the great majority of instances, the state either contains two or more distinctly different culture areas, or fragments thereof, or is a part of a much larger single culture area (for example, Iowa, North Dakota, Vermont). Thus there are sharp north-south dichotomies in Missouri, Illinois, Indiana, Ohio, and Florida; and Tennessee proudly advertises the fact that there are really three Tennessees. In one case, that of antebellum Virginia, the centrifugal cultural forces were so strong that actual fission took place in 1863 (into Virginia and West Virginia) along one of those rare interstate boundaries that approximates a genuine cultural divide.

Although geographers have devoted much attention to the structure of economic regions and the forces operating within them, we know very little about the anatomy or physiology of culture areas. The most interesting attempt to formulate general principles has been by Donald W. Meinig. He contends that, under ideal conditions (that is, given relative isolation over a long period, freedom from interference by others, and the opportunity for unhindered territorial expansion for a cultural group), three concentric zones will materialize. These he denotes, in order of decreasing degree of concentration or typicality, as the *core, domain,* and *sphere.* He defines them as follows:

A *core* area...is taken to mean a centralized zone of concentration, displaying the greatest density of occupance, intensity or organization, strength, and homogeneity of the particular features characteristic of the culture under study. It is the most vital center, the seat of power, the focus of circulation.

The *domain* refers to those areas in which the particular culture under study is *dominant,* but with markedly less intensity and complexity of development than in the core, where the bonds of connection are fewer and more tenuous and where regional peculiarities are clearly evident.

The *sphere* of a culture may be defined as the zone of outer influence and, often, peripheral acculturation, wherein that culture is represented only by certain of its elements or where its peoples reside as minorities among those of a different culture.[3]

The necessary conditions are almost ideally met in the Mormon culture region of the western United States (Figure 4.1) and in that instance, theory comfortably coexists with fact. More recently, Meinig has applied it successfully, though with modifications, to the case of

3 Donald W. Meinig, "The Mormon Culture Region: Strategies and Patterns in the Geography of the American West, 1847–1964," *Annals of the Association of American Geographers,* 55, No. 2 (1965), 213–17.

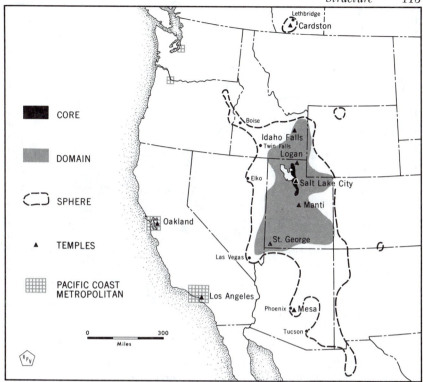

FIG. 4.1.   The Mormon culture region. (Adapted from: D. W. Meinig, "The
Mormon Culture Region," Annals of the Association of American
Geographers, 55, No. 2 [June 1965], p. 116.)

Texas[4] (Figure 4.2). There the domain has been subdivided into a
primary domain (which includes most of the state's territory) and a
secondary domain (the northern section of the Panhandle); and he has
also added a new fourth zone beyond the sphere—a "zone of penetration,"
evidently one of localized, "imperial" implantation of elements of Texan
culture, lying within portions of New Mexico, Arizona, Colorado, Utah,
and perhaps northern Mexico. What is most aberrant about the Texas
culture area is the geometry of its core, a bulbous, hollow triangle whose
three apices—the metropolises of Houston, Dallas-Fort Worth, and San
Antonio—have interacted complexly and have shared the functions usually confined to a single compact zone.

This simple scheme, it must be stressed, works readily only under
rather exceptional conditions and better at the national than the subnational scale. The growth and development of the great majority of
traditional culture areas, not only in the Old World but in North America
as well, has been inhibited by adjacent areas. The usual result is a

[4] Donald W. Meinig, *Imperial Texas: An Interpretive Essay in Cultural
Geography* (Austin: University of Texas Press, 1969), pp. 110–24.

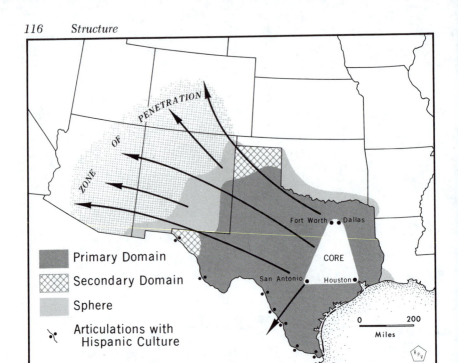

FIG. 4.2.   *The geographic morphology of the Texas empire. (Adapted from: D. W. Meinig,* Imperial Texas *[Austin: University of Texas Press, 1969], p. 93.)*

stunted, squashed, or misshappen area. The biological analogy is tempting: if you wish to see what an oak tree ought to look like ideally, find a mature specimen that has had a large meadow all to itself. Its brethren, struggling to survive in a dense grove, will have achieved very different sizes and shapes. Although the Meinig model accommodates the facts for the Mormon and Texas areas without much compromise, it can be made to work only with much stretching and bending in the New England, Midland, or French Canadian cases, and it is almost useless in describing or explaining other North American culture areas. It may be applicable at the national scale, with the Northeast (and especially Megalopolis) as the core; but the identification of domain, sphere or zone of penetration is a task still to be performed.

The cause and effect relationships between economic and culture areas in the United States and similar areas constitute another, largely unexplored subject. There is reason to believe that the two kinds of region were intimately associated during the colonial phase of our history, each taking form from and giving form to the other; but as time has gone on, the connections have become less clear. If one could, for example, formerly correlate New England (or at least its earlier, nodal segment) with a specific economic system, to do so is no longer feasible. It would appear that cultural systems are slower to respond to agents of change, whether endogenous or exogenous, than are economic or urban systems. Thus we have the spectacle of the American Manufacturing Belt—which happens to be a core region for many other social and economic activities than just manufacturing—now spanning major fractions of three traditional culture areas—New England, the Midland, and the Middle West—and the northern fringes of a fourth, the South. And its eastern facade and persistent nucleus, the great urbanoid sprawl known as Megalopolis, reaches from southern Maine to northern (or central?) Virginia, blithely ignoring the steep relict cultural slopes that are still visible in its more bucolic tracts.

The Major Traditional Culture Areas

AN HISTORICAL NOTE. The culture areas of the modern United States are European in origin, the result of importing European colonists and ways of life and the subsequent interactions among and within social groups and with a novel set of habitats. As suggested earlier, we can dismiss the aboriginal cultures as territorially insignificant, at least at the macro level.[5] Only in two subregions, the Southwest focused on the Upper Rio Grande and a portion of the Colorado Plateau (Figure 4.3) or the still indistinct Oklahoman subregion, does an Indian element merit consideration, and then only as one of several. Thus, except for the Southwestern Indian groups, pre-contact aboriginal areas can be safely ignored, but not without noting the curious coincidence or convergence of genetic factors in the case of western New York State. There the core area for the Iroquois Confederacy and its special near-civilization, which bloomed late in the pre-conquest period, matches perfectly a significant nineteenth-century culture hearth, which we are obliged to call, rather misleadingly for lack of a better alternative, the "Burned-Over District." In so far as North American aboriginal life attained a cultural climax outside the Mexican border zones, it was in the Southeast and Ohio Valley generally, and the lower Mississippi Valley most particularly. That area has no valid, latter-day Anglo-American descendant or counterpart. At the micro level, there are still many groups, some of mixed ancestry,[6] of local significance; but in almost every instance the culture in question has been greatly modified by European influences.

With some minor exceptions, the contemporary map of American culture areas (Figure 4.3) can be explained in terms of the genesis, development, and expansion of the three principal colonial culture hearths located along the Atlantic Seaboard.[7] Before taking up each in

[5] Delimited in Alfred L. Kroeber, *Cultural and Natural Areas of Native North America,* University of California Publications in American Archaeology and Ethnology, Vol. 38 (Berkeley: University of California Press, 1939).

[6] The geography of various triracial groups is presented in Edward T. Price, "A Geographic Analysis of White-Negro-Indian Racial Mixtures in Eastern United States," *Annals of the Association of American Geographers,* 43, No. 2 (1953), 138–55.

[7] Figure 4-3, which the reader will need to refer to repeatedly in the pages that follow, is apparently the first attempt ever made to delimit cultural areas for

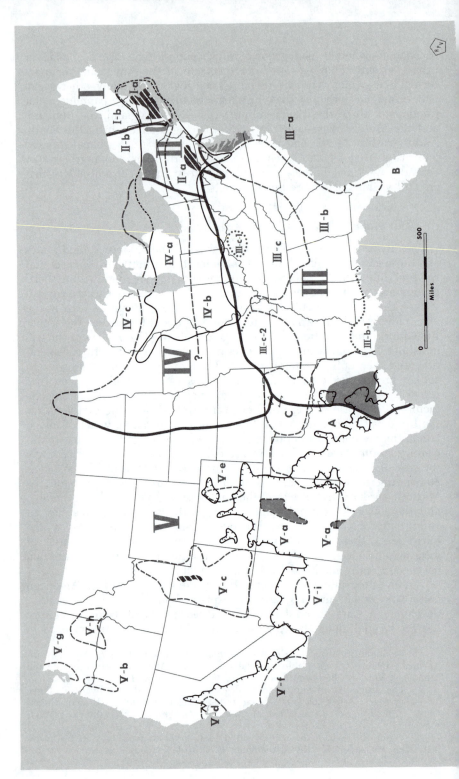

REGION	APPROXIMATE DATES OF SETTLEMENT AND FORMATION	MAJOR SOURCES OF CULTURE (listed in order of importance)
I. NEW ENGLAND	1620-1830	England
I-a. Nuclear New England	1620-1750	England
I-b. Northern New England	1750-1830	Nuclear New England, England
II. THE MIDLAND		
II-a. Pennsylvanian Region	1682-1850	England & Wales; Rhineland; Ulster; 19th Century Europe
II-b. New York Region, or New England Extended	1624-1830	Great Britain; New England; 19th Century Europe; Netherlands
III. THE SOUTH		
III-a. Early British Colonial South	1607-1750	England; Africa; British West Indies
III-b. Lowland, or Deep South	1700-1850	Great Britain; Africa; Midland; Early British Colonial South; aborigines
III-b-1. French Louisiana	1700-1760	France: Deep South; Africa; French West Indies
III-c. Upland South	1700-1850	Midland; Lowland South; Great Britain
III-c-1. The Bluegrass	1770-1800	Upland South; Lowland South
III-c-2. The Ozarks	1820-1860	Upland South; Lowland South; Lower Middle West
IV. THE MIDDLE WEST		
IV-a. Upper Middle West	1790-1880	New England Extended; New England; 19th Century Europe; British Canada
	1800-1880	Midland; Upland South; New England Extended; 19th Century Europe
IV-b. Lower Middle West	1790-1870	Upper Middle West; 19th Century Europe
IV-c. Cutover Area	1850-1900	Upper Middle West; 19th Century Europe

REGION	APPROXIMATE DATES OF SETTLEMENT AND FORMATION	MAJOR SOURCES OF CULTURE (listed in order of importance)
V. THE WEST		
V-a. Upper Rio Grande Valley	1590-	Mexico; Anglo-America; aborigines
V-b. Willamette Valley	1830-1900	Northeast U.S.
V-c. Mormon Region	1847-1890	Northeast U.S.; 19th Century Europe (Mexico)
V-d. Central California	(1775-1848) 1840-	Eastern U.S.; 19th Century Europe; Mexico; East Asia
V-e. Colorado Piedmont	1860-	Eastern U.S.; Mexico (Mexico)
V-f. Southern California	(1760-1848) 1880-	Eastern U.S.; 19th & 20th Century Europe; Mormon Region; Mexico; East Asia
V-g. Puget Sound	1870-	Eastern U.S.; 19th & 20th Century Europe; East Asia
V-h. Inland Empire	1880-	Eastern U.S.; 19th & 20th Century Europe
V-i. Central Arizona	1900-	Eastern U.S.; Southern California; Mexico
REGIONS OF UNCERTAIN STATUS OR AFFILIATION		
A. Texas	(1690-1836) 1821-	(Mexico) Lowland South; Upland South; Mexico; 19th Century Central Europe
B. Peninsular Florida	1880-	Northeast U.S.; the South; 20th Century Europe; Antilles
C. Oklahoma	1890-	Upland South; Lowland South; aborigines; Middle West

119

turn, two general comments may be in order. Even though each of the three is basically British in character, their personalities have been and still are quite distinct, thanks to different social and political conditions during the critical period of First Effective Settlement, along with localized physical and economic circumstances. Secondly, the cultural gradients tend to be much steeper and the boundaries between regions more distinct than is true for the remainder of the continent. There is greater variety within a narrower range of space as one approaches the older, inner regions, but greater blurring of cultural traits and confusion of identity in the reverse direction, toward the outer, newer tracts.

NEW ENGLAND. If we begin our inventory with the New England culture area, it is not because it is the oldest of our major cultural-geographic entities—an honor it shares with the South or perhaps the Hispanic Southwest—but rather because the region was the most clearly dominant during the century of national expansion following the American Revolution. New England may have been strong, if not preeminent, in the fields of manufacturing, commerce, finance, and maritime activity, but it was in the spheres of higher social and cultural life that the area exercised genuine leadership—in education, politics, theology, literature, science, architecture, and the more advanced forms of mechanical and social technology. Thus, if any single section must be nominated as the leading source of ideas and styles for the remainder of the nation from about 1780 to 1880, New England is the logical candidate.[8] Incidentally,

---

the entirety of the conterminous United States. An earlier map of more limited scope is Figure 1, "Rural-Farm Cultural Regions and Sub-Regions," in A. R. Mangus, *Rural Regions of the United States* (Washington, D.C.: Works Progress Administration, 1940), facing p. 1. Also of considerable interest to the cultural geographer is a map depicting the more or less subliminal regionalization of the country on the part of a test group of Ohio State University students, entitled "Locational Classes of Student Schemata of the U.S.A.," in Kevin R. Cox and Georgia Zannaras, "Designative Perceptions of Macro-Spaces: Concepts, a Methodology and Applications," in John Archea and Charles Eastman, eds., *Proceedings of the 2nd Annual Environmental Design Research Association Conference*, October 1970 (Pittsburgh: October 1970), p. 124.

The contention that the three colonial cultural hearths treated in this chapter are pivotal in the evolution of culture areas at the national scale has been advanced by a number of authors in the past, among them Carle C. Zimmerman and R. E. DuWors, *Graphic Regional Sociology (A Study in American Social Organization)* (Cambridge, Mass.: Phillips Book Store, 1952); Conrad Arensberg, "American Communities," *American Anthropologist*, 57, No. 6 (1955) 1143–60; Donald W. Meinig, "The American Colonial Era: A Geographic Commentary," *Proceedings of the Royal Geographical Society, South Australian Branch* (1957–1958), pp. 1–22; Hans Kurath, *A Word Geography of the Eastern United States* (Ann Arbor: University of Michigan Press, 1949); and Henry Glassie, *Pattern in the Material Folk Culture of the Eastern United States* (Philadelphia: University of Pennsylvania Press, 1968).

8 Or, as de Toqueville wrote perceptively, if a bit floridly, "The two or three main ideas which constituted the basis of the social theory of the United States were first combined in the Northern English colonies, more generally denominated the States of New England. The principles of New England spread at first to the neighboring States; they then passed successively to the more distant ones; and at length they imbued the whole Confederation. They now extend their influence

it also furnishes an impressive example of the capacity of strongly motivated communities for rising above the constraints of a parsimonious environment.

Relatively small though it is, the territory occupied by the six New England states may be usefully split into two subregions: the older, more densely settled, highly urbanized and industrialized southeastern segment; and the newer, rather thinly occupied northern area, taking in all of Vermont and nearly all of New Hampshire and Maine, an area colonized mostly after 1770 by settlers from the nuclear zone. It is the former tract, of course, which has been so potent as a source of people, ideas, and influences for much of North America. By virtue of location, wealth, and seniority, the Boston metropolitan area has acted as the functional hub for all of New England both culturally and economically; but its sovereignty is shared to some degree with two other old nuclear nodes: the lower Connecticut Valley and the Narragansett Bay region.

The demographic and ideological expansion of New England began quite early with the settlement of parts of Long Island, northern New Jersey, and the Hudson Valley. The thrust westward was so powerful and persistent that it is justifiable to call New York, northern New Jersey, northern Pennsylvania, and much of the Upper Middle West "New England extended." Nonetheless, strong though the New England component may have been in the Hudson Valley and much of upper New York State, the western political limits of Connecticut, Massachusetts, and Vermont (or, roughly, the line of the Taconics, Berkshires, and Lake Champlain-Lake George valley) do neatly demarcate an area that is not quite New England from one that truly is—that second and last good example we can muster of interstate boundaries being nearly identical with cultural zonation. The civil disturbances of the American Revolution stimulated a major exodus of loyalist New Englanders to Nova Scotia and Upper Canada (later Ontario), as well as to the British West Indies. The energetic endeavors of New England whalers, merchants, and missionaries resulted in a perceptible imprinting of population and culture in Hawaii, various other Pacific isles, and scattered points in the Caribbean. New Englanders, arriving by sea, were also active in the Americanization of early Oregon and Washington, with results visible to this day. Later, the overland diffusion of New England natives and

---

beyond its limits over the whole American world. The civilization of New England has been like a beacon lit upon a hill, which, after it has diffused its warmth around, tinges the distant horizon with its glow." Alexis de Toqueville, *Democracy in America* (New York: Oxford University Press, 1947), p. 33. The primacy of New England in higher cultural development is also strongly emphasized in Hugo Münsterberg, *The Americans* (New York: McClure, Phillips & Co., 1905), pp. 347–64 and Daniel J. Boorstin, *The Americans: The National Experience* (New York: Random House, Inc., 1965), pp. 3–48. The most useful and convenient general guide to the social and cultural characteristics of the region is Hans Kurath, *Handbook of the Linguistic Geography of New England* (Providence, R.I.: Brown University, 1943). The topic of demographic and cultural expansion is handled in Lois (Kimball) Mathews Rosenberry, *The Expansion of New England: The Spread of New England Settlement and Institutions to the Mississippi River, 1820–1865* (Boston: Houghton Mifflin Company, 1909).

practices meant a recognizable New England flavor not only for the Upper Midwest from Ohio to the Dakotas but also more dilutely in the Pacific Northwest in general. Beyond the immediate impact of businessman and settler, there is the dissemination of less palpable items—ideas, attitudes, and innovations of all sorts—perhaps of greater weight ultimately than the economic or demographic factors. In keeping with the basic criterion for defining a culture area, New Englanders have been keenly, happily aware of their identity, though probably more so in the past than at present; and they have usually been readily labeled by outsiders by virtue of pattern of speech, religion, behavior, or thought. The massive immigration of persons from Ireland, Germany, then eastern and southern Europe and Quebec that began in the 1830's has brought about important shifts in the character of New England society, but the sense of its special early individuality lingers on.

THE SOUTH. The same observation applies with even greater force to natives of the South, by far the largest of the three primordial Anglo-American culture areas and also the most aberrant of any with respect to national norms.[9] Indeed the South has been so distinct from the non-South in almost every observable or quantifiable feature and so fiercely jealous of its peculiarities that for some years the question of whether it could maintain political and social unity with the non-South was in serious doubt. Only during the twentieth century can one argue for a decisive convergence with the rest of the nation, at least in economic behavior and material culture. An early, persistent deviation from the national mainstream probably commenced during the first years of settlement. The Caucasoid settlers of the South were almost purely British and not outwardly different from the British, Welsh, or Scotch-Irish who flocked to New England or the Midland; but almost certainly they were different in motives and social values. The immigration of African slaves may have been a factor, as might also a degree of interaction with the aborigines that was missing further north. And certainly the unusual (for Northwest Europeans) pattern of economy, settlement, and social organization, in part a matter of a starkly unfamiliar physical habitat, may have prompted a swerving away from other culture areas.

In both origin and spatial structure, the South has been characterized by diffuseness. If we must narrow the search for a nuclear hearth down to a single tract, the most plausible choice is the Chesapeake Bay area and the northeastern corner of North Carolina, the earliest area of recognizably Southern character. Unlike the nodal zones for New England or the Midland, this was, and still is, an emphatically rural area, despite the growth of two large metropolitan districts—Baltimore and the Hampton Roads conurbations—toward either end of the Bay. It was via this corridor and the routes running further inland from the Delaware River ports that the great bulk of the Southern territory was eventually peopled.

9 For general introductions to the cultural geography of the South, consult Rupert B. Vance, *The Human Geography of the South* (Chapel Hill: University of North Carolina, 1935) and J. Fraser Hart, *The Southeastern United States* (New York: Van Nostrand Reinhold Company, 1967).

But early components of Southern population and culture also arrived from other sources via other routes. A narrow coastal strip running from Carolina down to the Georgia-Florida boundary, and including the Sea Islands, decidedly Southern in flavor yet self-consciously standing apart from other parts of the South, was colonized directly from Great Britain. There were also significant direct connections with the West Indies; and it is here that the African cultural contribution is strongest and purest. The cities of Charleston and Savannah, which have nurtured their own quite special civilizations, dominated this subregion in every sense. Similarly, we can demarcate a French Louisiana that received elements of culture and population, to be stirred into the special Creole mixture, not only from a putative Chesapeake Bay hearth area but also directly from France, Acadia (Colonial French Nova Scotia), the French West Indies, and, of course, Africa. In the south central Texas, the Teutonic influx was so heavy that a special subregion can quite properly be designated.

It would seem, then, that the Southern culture area may be an example of convergent or parallel evolution of a variety of elements arriving along several paths, but subject to some single general process that could mold one larger regional consciousness and way of life. This in contrast to the more common experience of colonists and cultural forces radiating outward from a single early source region. The result, in any case, is a very extensive territory without any truly coherent structure, à la the Meinig Model, or any other that can be suggested. Where was the focal center of the Confederacy: Richmond, Montgomery, Atlanta, or somewhere else? Simply to articulate the question is to realize how difficult it is. One can point to the Cotton Belt, an economic region of variable dimensions, stretching in a great crescent from south central Virginia to east Texas, as the "modal South," one that contains in heightened form nearly all the traits regarded as typically Southern; but it is impossible to claim that the South's historic roots lie there or that it ever operated as an organizing center, economically or culturally. The real power and fount of ideas resided elsewhere.

The South can be subdivided into a much greater number of subregions than can any of the other, older traditional regions. Three have already been mentioned; but these are of a lesser order than the two principal Souths, variously called Upper and Lower South, Upland and Lowland South, or Yeoman and Plantation South.[10] The former, which comprises the Southern Appalachians, the upper Appalachian Piedmont,

---

[10] This basic dichotomy is made, verbally and cartographically, in Wilbur Zelinsky, "Where the South Begins: The Northern Limits of the Cis-Appalachian South in Terms of Settlement Landscape," *Social Forces*, 30 (1951), 172–78; Fred Kniffen, "Folk Housing: Key to Diffusion," *Annals of the Association of American Geographers*, 55, No. 4 (1965), 571–73; Terry G. Jordan, "The Imprint of the Upper and Lower South on Mid-Nineteenth Century Texas," *Ibid.*, 57, No. 4 (1967), 667–90; Hart, *The Southeastern United States;* and Glassie, *Pattern in the Material Folk Culture*, p. 39. The most detailed attempt at subdividing the South—or the larger part of it—into cultural (in this instance, dialect) regions is offered in Gordon R. Wood, *Vocabulary Change: A Study of Variation in Regional Words in Eight of the Southern States* (Carbondale: Southern Illinois University Press, 1971), p. 358.

the Cumberland and other low interior plateaus, and the Ozarks and Ouachitas, was occupied, both culturally and demographically, from two sources: the Chesapeake Bay hearth area and the early Midland. The latter area includes the greater part of the South Atlantic and Gulf coastal plains, along with the lower Appalachian Piedmont, and originated principally in the Chesapeake Bay area, with only minor influences from the coastal Carolina and Georgia belt, Louisiana, or elsewhere. The division between the two subregions remains distinct all the way from Virginia to Texas.[11] Within the Upland South, the Ozark segment might legitimately be detached from the Appalachian;[12] and within the latter, the Kentucky Bluegrass region certainly merits special recognition.

It is toward the margins of the South that one encounters the greatest difficulties in delimiting and assigning subregions. The outer limits themselves are a topic of special interest.[13] There seems to be more than an accidental relationship between these limits and various aspects of the North American climate. Thus the northern boundary, a fuzzy gradational zone, but one definitely not related to the conventional Mason and Dixon Line or the Ohio River, seems isothermal in character, the association being with length of frost-free season or with temperature during the winter months. As the Southern cultural complex was carried to the west, it not only retained its strength but became even more intense, in contrast to the experience of New England or the Midland. But the South finally fades away as one approaches the famous 100th Meridian and a critical decline in annual precipitation. The apparent correlation of the cultural South with a humid subtropical climate is a provocative one.

A possible emergent Oklahoman subregion is still too indistinct to present any immediate problem; but the Texas subregion is so large, unmistakable, vigorous, and self-assertive that it presents some vexing taxonomic and cartographic questions. Meinig has expertly discussed the varied outer influences that have helped make Texas the special place culturally that it undoubtedly is;[14] and he has gone further by subdividing it into nine component areas in a suggestive map. (See Figure 4.4.) But he leaves unresolved the larger matter of whether Texas is simply a subordinate fraction of the Greater South or whether it has now acquired so strong and divergent an identity that it can be denoted a new first-order region. We have chosen to equivocate in Figure 4.3 by showing the Texan region as straddling the frontier of the regular South. It is conceivable that we are witnessing the genesis of a major region in a frontier zone where several distinct cultural communities confront each other, as per the hypothesis put forward by Leslie White.[15]

11 Terry G. Jordan, "The Texas Appalachia," *Annals of the Association of American Geographers,* 60, No. 3 (1970), 409–27.

12 E. Joan Wilson Miller, "The Ozark Culture Region as Revealed by Traditional Materials," *Annals of the Association of American Geographers,* 58, No. 1 (1968), 51–77.

13 Zelinsky, "Where the South Begins."

14 Donald W. Meinig, *Imperial Texas.*

15 Leslie A. White, *The Science of Culture* (New York: Farrar, Straus & Giroux, 1949).

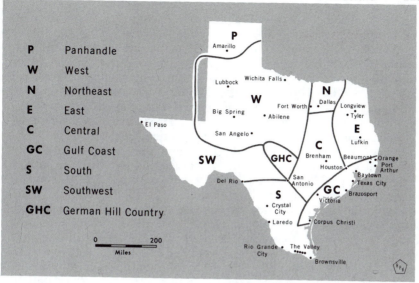

P	Panhandle
W	West
N	Northeast
E	East
C	Central
GC	Gulf Coast
S	South
SW	Southwest
GHC	German Hill Country

FIG. 4.4. The culture areas of Texas in the 1960's. (Adapted from: D. W. Meinig, Imperial Texas [Austin: University of Texas Press, 1969], p. 93.)

In a few more decades we may be able to decide. Similarly, peninsular Florida is either within or juxtaposed to the South, but without being unequivocally part of it. In this case, an almost empty territory began to receive significant settlement only after about 1890; and, if like Texas, most of it came from the older South, there were also vigorous infusions from elsewhere. We shall have occasion to return to the identity problems of both peninsular Florida and Texas at a later juncture.

THE MIDLAND. This region has been reserved for last among the major colonial entities not because its significance has been slighter, but because it is the least conspicuous, either to outsiders or to its own inhabitants. And indeed this fact may reflect its centrality to the course of American development. The serious European occupation and development of the Midland began a generation or more after that of the other major cultural nodes, and after several earlier, relatively ineffectual trials by the Dutch, Swedes, Finns, and British. But, once begun late in the seventeenth century by William Penn and associates, the colonization of the area was an instant success. This was especially so in the lower Delaware and Susquehanna valleys; and Philadelphia, the major port of entry, also became preeminent as economic, social, and cultural capital. It was within southeastern Pennsylvania that this culture area first assumed its distinctive form: a prosperous agricultural society, then quite quickly a mixed economy as mercantile and later industrial functions came to the fore. By the middle of the eighteenth century, much of the region has acquired a markedly urban character, resembling in many ways the more advanced portions of the North Sea countries. In this respect at least, the Midland was well ahead of neighboring areas to the north and south.

It also differed in its polyglot ethnicity. From almost the very beginning, the full range of ethnic and religious groups of the British Isles were joined by immigrants from the European mainland. This diversity has, if anything, grown through the years—and promises to persevere indefinitely. The spatial mosaic of colonial ethnic groups has persisted in much of Pennsylvania, New York, New Jersey, and Maryland, as has the remarkable variety of more recent nationalities and churches in coal fields, company towns, cities large and small, and many a rural tract. Much the same sort of ethnic heterogeneity is to be seen in New England, the Middle West, and a few other areas, but the Midland still stands out as our most polyglot region.

If we limit our attention to Pennsylvania and the adjacent tracts of New Jersey, Delaware, and Maryland, a useful distinction can be made (though poorly noted in the literature) between an inner zone, which I call the Pennsylvanian subregion for lack of a better term, in which the Midland culture reached its climactic development, and an outer zone, which is much more nondescript. The former is entirely inland and focused, insofar as it has any core area, upon those sections of the upper Piedmont and the Great Valley lying to the west and northwest of Philadelphia. Although the metropolis is physically separated from the Pennsylvanian subregion, there is every reason to infer a strong umbilical connection during the formative period. It is in the Pennsylvania subregion that such conspicuous diagnostic features as the forebay barn, a localized town morphology, and various house types are located, along with other cultural elements that are less well advertised.[16] The curious fact that the quintessential Midland should have crystallized and flourished well inland rather than within the immediate Philadelphia hinterland is not readily explainable and deserves further study. That this circum-Philadelphia area, in much of Bucks, Montgomery, Chester, and Delaware counties, is sharply set off from the Pennsylvania subregion in terms of ethnic composition, settlement morphology, architecture, and probably other ways is easily established. It is almost totally British, with an especially strong representation of the Welsh and the Quakers. Within the Pennsylvania subregion, the Teutonic element has always been notably strong, accounting for more than 70 per cent of the population of many townships. If it were not for the fact that the Anglo-American culture finally proved supreme, one would indeed be tempted

16 Richard Pillsbury, "The Urban Street Pattern as a Culture Indicator: Pennsylvania, 1682–1815," *Annals of the Association of American Geographers*, 60, No. 3 (1970), 428–46 and Joseph W. Glass, *The Pennsylvania Culture Region: A Geographic Interpretation of Barns and Farmhouses*, doctoral dissertation, Pennsylvania State University, 1971). The best general overview of the Midland is probably still Thomas Jefferson Wertenbaker, *The Founding of American Civilization: The Middle Colonies* (New York: Charles Scribner's Sons, 1938). For an interesting comparison of the regional personality of the Pennsylvania German area with that of New England, see Frank R. Kramer, *Voices in the Valley: Mythmaking and Folk Belief in the Shaping of the Middle West* (Madison: University of Wisconsin Press, 1964), pp. 87–104.

to designate the area as Pennsylvania German. Considerations of land-forms and migrational history have carried the subregion into the Maryland Piedmont; and although its width tapers quickly beyond the Potomac, it does reach into parts of Virginia and West Virginia. As intimated earlier, the demographic and cultural effects of the Midland and the Pennsylvania subregion are legible far down the Appalachian zone and into the South in general. Certain sections of central North Carolina, for example, seem almost like detached fragments of the mother region, as evidenced by a highly un-Southern religious complexion, among other things. Back within Pennsylvania, the areas to the west and north of the two subregions already described are less clearly Midland in character; and much of central and southern New Jersey, except perhaps for a zone within a few miles of Philadelphia, is difficult to classify in any way, and is assigned to the Midland largely by default.

The northern half of the Midland Region, the New York subregion, or New England Extended, presents the cultural geographer with a major classificatory dilemma. Perhaps it is sensible to think of it as a hybrid place formed mainly from two parental strains of almost equal potency: New England and the post-1660 British element moving up the Hudson Valley and beyond. In addition, there has been a persistent, if slight, residue of early Dutch culture and some subtle filtering north-ward of Pennsylvanian influences. Evidently, it is within the New York subregion that we find the first major *intra*-American blending and fusion of regional cultures, most particularly within the early nineteenth-century Burned-Over District in and near the Finger Lakes and Genesee areas. This locality, another of the many about which we still know much too little, is notable in two respects. It was the seedbed for a number of important innovations;[17] and it was also a major staging area for westward migration and quite possibly a chief source for the people and notions that were to build the Middle Western Region. The boundary between the New York subregion and the (Greater) Pennsylvania sub-region in northern Pennsylvania and New Jersey can be delineated rather sharply and in a number of different ways, including language, religion, house types, and town morphology. It also coincides with a significant physiographic barrier, the higher portions of the Allegheny Plateau, and the southern limit of late Wisconsin glaciation. Apparently migrants were deflected from both sides of this border zone by the lay of the land, drainage patterns, and generally poor soil quality.

Toward the west, the Midland or, more specifically, its Pennsylvania subregion, retains its integrity only for a short distance, certainly no further than eastern Ohio, as it pinches out between the South and New England Extended or becomes submerged within the Middle West. Still

---

17 This is clearly the case for innovations in toponymic practice, as demonstrated in Wilbur Zelinsky, "Classical Town Names in the United States: The Historical Geography of an American Idea," *Geographical Review*, 57, No. 4 (1967), 463–95. The role of west central New York in nineteenth-century religious developments is detailed in Whitney R. Cross, *The Burned-Over District: The Social and Intellectual History of Enthusiastic Religion in Western New York, 1800–1850* (Ithaca, N.Y.: Cornell University Press, 1950).

the significance of the Midland in the genesis of the Middle West or of the national culture must not be belittled. The precise details remain to be worked out, but it is probable that the Midland contribution to overall national patterns will prove to be substantial, just as with the other two colonial hearth areas. Its very success in projecting its image upon so much of the remainder of the country may have rendered the source area less visible. As both the name and location would suggest, the Midland is intermediate in character in many respects between New England and the South. Moreover, its residents are much less concerned with, or conscious of, its existence (excepting the "Pennsylvania Dutch" caricatures) than is true for the other regions, and the Midland also happens to lack their strong political and literary traditions.[18]

THE MIDDLE WEST. There is no such self-effacement in the case of the Middle West, that large triangular region which is justly regarded as the most modal, the section most nearly representative of the national average. Everyone within or outside the Middle West knows of its existence, but no one seems sure where it begins or ends.[19] The older apex of the eastward-pointing equilateral triangle appears to rest in the vicinity of Pittsburgh, and the two western corners melt away somewhere in the Great Plains, possibly southern Manitoba and southern Kansas, respectively. Even though the national sections are distinct enough, this is a clear case of a North American culture area transcending an international boundary.[20] Both southern Ontario and southern Manitoba display an affinity for the adjacent American area that rivals their cultural allegiance to other portions of Canada. The boundary question is irrelevant only along the northeastern side of the Midwest as it gives way to bleak pre-Cambrian country. The eastern terminus and the southern and western borders are obviously broad, indistinct transitional zones.

The historical geography of the Middle West remains largely unstudied, but we can plausibly conjecture that this culture region must be the progeny of all three Colonial regions and that the fertile union took

18 Has Pennsylvania produced any stateman of truly national stature, aside from Benjamin Franklin, that prototypical American, who happens to have been born and reared in Boston? And John O'Hara would seem to be the most eminent author having a strong identification with the Commonwealth. The question of this relatively modest political or literary achievement, at least since Revolutionary days, among a technologically advanced and affluent population, and the contrast in this respect with New England and the South, is a tantalizing one.

19 Apparently the only serious effort thus far to map the limits of the Middle West is that in Joseph W. Brownell, "The Cultural Midwest," *Journal of Geography*, 59 (1960), 43–61.

20 The geometry of Canadian culture areas contrasts strongly with that of the United States. In the latter, the cultural "grain" runs from east to west, while that of Canada would seem to trend from south to north. Moreover, each of the five major Canadian regions—the Maritimes, French Canada, upper Canada (portions of Ontario and southwest Quebec), the Prairie Provinces, and British Columbia—is isolated both culturally and physically from the others, with firmer linkages to its neighboring United States culture area (except for French Canada) than with its Canadian confrères.

place in the upper Ohio Valley.[21] The early routes of travel, employing the Ohio and its tributaries, the Great Lakes, and the low, level corridor along the Mohawk and the Lake Ontario and Lake Erie coastal plains, all converge upon the state of Ohio. It was there that the people and cultural traits from New England, the Midland, and the South were first funneled together. Thence there would seem to have been a fan-like widening westward of the new hybrid area as pioneer settlers worked their way frontierward. Like the South, the Middle West lacks a genuine focal zone or city around which both ideas and commerce were built (although an unconvincing case might be improvised for Chicago). A parallel to the Cotton Belt's role as a climactic zone of development is the central portion of the Corn Belt, or the Cash Grain Region; but it is plainly not a core area in any functional sense.

Two major subregions are readily discerned: the upper and lower Middle West. They are separated by a line roughly approximating the 41st Parallel, one that persists as far west as Colorado in terms of speech patterns[22] and indicates regional differences in ethnic and religious terms as well. Much of the upper Middle West retains a faint New England bouquet, although the Midland is probably equally important as a source of influences. A rich mixture of German, Scandinavian, Slavic, and other non-WASP elements has greatly diversified a stock in which the British element is usually dominant, and the range of church denominations is great and varied. The lower Middle West tends to resemble the South, except for the relative scarcity of Negroes, in its predominantly Protestant and British makeup. There are, of course, many local exceptions, areas of Catholic and non-WASP strength; but on the whole the subregion tends to be more nativistic in inclination than other parts of the nation.

THE PROBLEM OF THE "WEST." The culture areas so far discussed account for roughly the eastern half of the conterminous United States. One faces a genuine dilemma in attempting to arrive at an adequate cultural classification for the remaining half. The concept of an "American West" is strong in the popular imagination and is constantly reinforced by the romanticized images of the cowboy genre. It is tempting to succumb and accept the widespread Western livestock complex as somehow epitomizing the full gamut of Western life, but this would be intellectually dishonest. The cattle industry may have accounted for more than half the active Western *ecumene*, as measured in acres, but it accounts for only a relatively small fraction of the total population; and in any case a single sub-culture cannot represent the total regional culture.

Does a genuine, single, grand Western culture region exist? We cannot be certain until the relevant evidence has been rigorously ex-

---

[21] The question is dealt with, but in tangential fashion, in Kramer, *Voices in the Valley.*

[22] Clyde T. Hankey, *A Colorado Word Geography*, Publication of the American Dialect Society, No. 34 (University, Ala.: University of Alabama Press, 1960); and Roger W. Shuy, *The Northern-Midland Dialect Boundary in Illinois*, Publication of the American Dialect Society, No. 38 (University, Ala.: University of Alabama Press, 1962).

amined. This has not been done to date, nor will it be easy. When suitable measures are devised, it is possible that the cultural distance between, say, San Diego and Fort Worth or Pocatello will appear to be at least as great as that between Kankakee and Harrisburg. Unlike the East, where settlement is virtually continuous through space, and culture areas and subregions abut and overlap in splendid confusion, the west features eight major (and many lesser) nodes of population, separated from each other by wide expanses of nearly uninhabited mountain or arid desert. The only two obvious properties all these isolated clusters have in common is the recent intermixture of several strains of culture, primarily from the East, but with additions from Europe, Mexico, and East Asia as well, and (except for one) a general modernity, having been settled in a serious way no earlier than the 1840's. I should prefer to consider some as inchoate, partially formed cultural entities. The others have acquired definite personalities, but are difficult to classify as first-order or lesser-order culture areas at this time. Their grouping together on Figure 4.3 is a cartographic convenience, not a matter of intellectual conviction.

There are three, and possibly four, major tracts in the western United States that reveal a genuine cultural identity: the upper Rio Grande region; the Mormon region; Southern California; and central California. (To this group one might also add the anomalous Texan and Oklahoman subregions, which may adhere to either the South or the "West," or both). The term upper Rio Grande region has been coined to denote the oldest and strongest of the three sectors of Hispanic-American activity in the American Southwest, the other two having been Southern California and portions of Texas. Although the region is focused upon the valley of the upper Rio Grande, we can also include with it a broader domain taking in segments of Arizona and Colorado as well as other parts of New Mexico. European communities and culture have been present in strength, with only one interruption, since the late sixteenth century. The initial sources were Spain and Mexico, but after 1848 at least three distinct strains of Anglo-American culture have been increasingly well-represented: the Southern, Mormon, and a general undifferentiated northeastern American culture, along with a distinct Texan subcategory.[23] But, for once, all this has occurred without obliterating the aboriginal folk, whose culture endures in various stages of dilution, from the strongly Americanized or Hispanicized down to the almost undisturbed. The general pattern is that of a mosaic, with the aboriginal, Anglo, and Hispanic constituting the main elements. Furthermore, all three major groups, not just the Anglo-American, are complex in character. The aboriginal component embraces the Navaho, Pueblos, and several smaller groups, each of which is quite distinct from the others. The

[23] John L. Landgraf, *Land-Use in the Ramah Area of New Mexico: An Anthropological Approach to Areal Study*, Papers of the Peabody Museum of American Archaeology and Ethnology, Vol. 42, No. 1 (Cambridge, Mass.: Harvard University, 1954); and Donald W. Meinig, *Southwest: Three Peoples in Geographical Change, 1600–1970* (New York: Oxford University Press, 1971).

"Hispanic" element is also diverse—modally Mexican Mestizo, that is, Euro-American, but ranging from pure Spanish to nearly pure pre-Spanish aboriginal. Given the strong recent influx of Mexicans and Anglos (many with a cultivated taste for the distinctiveness of the area) and its general economic prosperity, this is one region that appears destined to wax stronger and to nurture zealously its special traits.

The Mormon region also promises to maintain its strength and cultural separateness for many years to come, for it is emphatically expansive in religion and demography, even though it has ceased to expand territorially, as it did so vigorously in the first few decades after establishment.[24] Despite its Great Basin location and an exemplary adaptation to environmental constraints, this cultural complex still appears somewhat non-Western in spirit: The Mormons may be in the West, but they are not entirely of it. The historical derivation from the Middle West and, beyond that, from ultimate sources in New York and New England, is still apparent, along with the generous admixture of European converts. Here again, as in New England, the power of the human will and an intensely cherished abstract design have triumphed over an unfriendly habitat. Despite its relative recency and an unflagging vigor, the region has functioned, at least since the great ingathering of the Latter Day Saints during the past century, as a traditional region—one inhabited by a (religiously defined) tribal group, with a deep natal attachment to a particular tract, and almost ideally structured spatially, as Meinig has shown, in accordance with the principles one would postulate theoretically for an isolated culture area. The Mormon way of life is expressed in many recognizable ways in the settlement landscape and economic activities within a region that is more homogeneous internally than any other American culture area.[25] This is also the only large portion of the United States, aside from the South, whose political allegiance to the federal Union was ever seriously in doubt.

In contrast, the almost precisely coeval Central California region has not yet gained its own strong cultural coloration, except possibly for the city of San Francisco and some of its suburbs. From the first weeks of the great Gold Rush onward, the area drew a thoroughly diverse population from Europe and Asia as well as the older portions of America. Speaking impressionistically, as one must for most Western regions, it is not yet obvious whether the greater part of the Central California region has produced a local culture amounting to more than the averaging out of the contributions brought by in-migrants, early and late. San

---

[24] Donald W. Meinig, "The Mormon Culture Region: Strategies and Patterns in the Geography of the American West, 1847–1964," *Annals of the Association of American Geographers*, 55, No. 2 (1965), 191–220.

[25] Joseph E. Spencer, "The Development of Agricultural Villages in Southern Utah," *Agricultural History*, 14 (1940), 181–89 and "House Types of Southern Utah," *Geographical Review*, 35 (1945), 444–57; Lowry Nelson, *The Mormon Village: A Pattern and Technique of Land Settlement* (Salt Lake City: University of Utah Press, 1952); and Richard V. Francaviglia, "The Mormon Landscape: Definition of an Image in the American West," *Proceedings of the Association of American Geographers*, 2 (1970), 59–61.

Francisco, the regional metropolis, may, however, have crossed the qualitative threshold. An unusually cosmopolitan outlook, including an awareness of the Orient stronger than that of any other American city, a fierce self-esteem, and a unique townscape may be symptomatic of a genuinely new, emergent local culture. In any case, the Central California region, along with the Southern Californian and the Upper Rio Grande, has recently acted as a creative zone for the nation as a whole, the springboard for many a fresh innovation.

The Southern California region is the most spectacular of the Western regions, not only in terms of economic and population growth, but also for the luxuriance, regional particularism, and general avante-garde character of its swiftly evolving cultural pattern.[26] Until the coming of a direct transcontinental rail connection in 1885, Southern California was remote, rural, and largely inconsequential. Since then, the invasion by persons from virtually every corner of North America and by the foreign-born has been massive and ceaseless; and a loosely articulated series of conurbations have been encroaching upon what little is left of arable or habitable land in the coast ranges and valleys from Santa Barbara to the Mexican border. Although every significant ethnic and racial group and every other American culture area is amply represented, there is reason to suspect (again largely on subjective grounds) that a process of selection for certain kinds of people, attitudes, and personality traits may have been at work at both source and destination. Certainly the region is aberrant from, or in the vanguard of, the remainder of the nation. One might view Southern California as the super-American region or the outpost of a rapidly approaching postindustrial future; but in any event its cultural distinctiveness, in visible landscape and social behavior, is evident to all. It is also clear that in Southern California we do not behold anything approaching a traditional region, or even the smudged facsimile of such, as is to be found in the eastern United States, but rather the largest, boldest experiment anywhere in creating a voluntary region. In discussing this new variety of geographic phenomenon, we shall shortly return to the Southern California scene.

The remaining Western regions identified in Figure 4.3—the Willamette Valley, Puget Sound, Inland Empire, and Colorado Piedmont—can be treated jointly as potential or emergent culture areas still too close to the national mean to arouse much self-awareness among their inhabitants or special curiosity on the part of the outsider. Conditions may have been propitious for such a regional blossoming in the Inland Empire a century ago or so, according to Meinig, but the unfortunate trisection of the area by Washington, Oregon, and Idaho, through the vagaries of boundary drawing, lessened the chances for such a

26 William L. Thomas, Jr., ed., "Man, Time, and Space in Southern California: A Symposium," *Annals of the Association of American Geographers*, Vol. 49, No. 3, Part 2 (1959); Carey McWilliams, *California: The Great Exception* (Westport, Conn: Greenwood Press, 1949); and Dennis Hale and Jonathan Eisen, eds., *The California Dream* (New York: The MacMillan Company, 1968).

development.[27] In all four regions (and also in the analagous case of British Columbia), one can witness the arrival of a cross-section of the national population and the growth of regional life around one or more major metropolises. A New England element has, however, been noteworthy in the Williamette Valley and Puget Sound regions, whereas a Hispanic-American component appears in the Colorado Piedmont. Only time and further study will reveal whether any of these regions, so distant from the historic sources of American population and culture, can engender the capacity or will for an independent cultural existence.

LESSER-ORDER CULTURE AREAS. Thus far we have scanned the top three levels of the hierarchy of culture areas within the United States: the nation itself; the five or more first-order culture areas; and a number of second-order culture areas. But, contrary to the popular myth of cultural homogeneity, there also exists a multitude of much smaller tracts, each distinctive as to culture. These microregions are both traditional and voluntary in type; and if we have only a sparse scattering of serious descriptions of the former, there is virtually no literature on the latter.

The most conspicuous of these small, culturally distinctive areas are several cities that maintain intimate economic and social connections with their hinterlands, yet have personalities all their own. The case of San Francisco has been noted already, and we can add to it such places as Charleston, Savannah, New Orleans, St. Louis, Philadelphia, Baltimore, Cincinnati, Detroit, Chicago, and New York—each a world unto itself, despite a functional centrality to a much broader area. This uniqueness is attested to by the unique speech patterns, at least in New York City, Boston, Philadelphia, and Chicago;[28] and some observers would testify that linguistic peculiarities are recognizable at the borough or neighborhood level.

The persistence of very minute culture areas in both urban and rural areas has been reported by the few geographers and others who have troubled to search for them. These dwarfish entities are most clearly visible in terms of ethnicity or religion. They have been described in Kalamazoo and Denver,[29] and definitively in Chicago's Near West Side.[30] Within a rural setting, an almost ideal example is to be seen in the Mennonite culture of central Pennsylvania's Kishicoquillas Valley. The distinctive character of yet another small valley, in northern New Jersey,

[27] Donald W. Meinig, *The Great Columbia Plain: A Historical Geography, 1805–1910* (Seattle: University of Washington Press, 1968), pp. 359–64.

[28] Metropolitan New York and Philadelphia are accorded subregional status in a map delimiting "The Speech Areas of the Eastern States," Kurath, *A Word Geography of the Eastern United States*, Figure 3.

[29] John A. Jakle and James O. Wheeler, "The Changing Residential Structure of the Dutch Population in Kalamazoo, Michigan," *Annals of the Association of American Geographers*, 59, No. 3 (1969), 441–60; and Daniel F. Doeppers, "The Globeville Neighborhood in Denver," *Geographical Review*, 57, No. 4 (1967), 506–22.

[30] Gerald D. Suttles, *The Social Order of the Slum: Ethnicity and Territory in the Inner City* (Chicago: University of Chicago Press, 1968).

has been analyzed in detail by Peter Wacker.[31] The great profusion of rural ethnic neighborhoods has been precisely documented for the entire state of Kansas in an exemplary atlas.[32] But clearly there remains an imposing agenda of unfinished business for the cultural and social geographer working at the microscopic level in Anglo-America.

The Voluntary Region

A new geometry of cultural space may have begun to materialize in the United States. The series of traditional spatial parcels, that is, discrete, contiguous patches of real estate, within each of which relative uniformity prevails, may be in the process of being replaced by a multi-layered sandwich (or hyperspace) in which very numerous strata of variable thickness tend to span the entire country. Formerly if one sought cultural diversity, one would move laterally or horizontally along the earth's surface from one place-bound group to another. Now, with minimal physical effort, one may experience this variety by shifting vertically, so to speak, in cultural space, from one layer of the sandwich to another— or by hobnobbing with different subcultures vicariously by means of telecommunications, which have brought about truly revolutionary changes in the physical, social, and mental mobility of individuals. To pursue our metaphor of the recent up-ending of interfaces between cultural groups, a particular layer of the cultural sandwich, (a subculture) may thicken so markedly at one point or another in its transcontinental expanse that that group then dominates the locality. When this occurs, for whatever reason, a voluntary region has come into being. Let us search for a few.

Voluntary regions are so novel and recent that few observers have consciously recognized or described them. Much of this discussion, then, must remain speculative, based on scattered observations, hints, and hunches. The massive American economic, social, and demographic data gathering systems are not yet attuned to their existence. This lag is illustrated by the relative dearth of hard data on tourism, recreation, and leisure time activities. Although these activities are of much concern to the student of voluntary regions and also to many businessmen, we are not yet reconciled to the pagan notion that these are serious items, socially or academically. There is also the matter of scale, whether spatial or temporal. Most of the emergent regions are tiny, fragmentary, or ephemeral. But these are inadequate reasons for ignoring them, for the smallness or spatial irregularity of the voluntary region may simply reflect the new spatial geometry of post-industrial communities, about which very little is yet known. Furthermore, all spatial structures, not just the cultural, are dynamic and evanescent. The spatially coherent cultural group that coheres for five days or five hours is just as real, and perhaps

---

31 Peter O. Wacker, *The Musconetcong Valley of New Jersey: A Historical Geography* (New Brunswick, N.J.: Rutgers University Press, 1968).
32 J. Neale Carman, *Foreign-Language Units of Kansas. I. Historical Atlas and Statistics* (Lawrence: University of Kansas Press, 1962).

as interesting, as one that persists for five hundred or five thousand years.

There are, however, at least two extensive tracts within the United States that appear to be taking on the character of voluntary regions and show every sign of relative stability: the Southern Californian and the peninsular Florida subregions. (A rather weaker case might also be made out for the upper Rio Grande and one or two other Western subregions.) The former two began, conventionally enough, as agricultural and ranching frontiers, with the regular later accretion of higher economic functions. Then, more recently, it has become evident that the large volume and character of the continuing in-migrations were controlled by factors other than economic. The opportunity to prosper remains important, but more and more of those bound for portions of Southern California and Florida seem to be seeking congenial surroundings and like-minded companions, even though we cannot yet specify the precise ways in which these perceptions, decisions, and migrations have been occurring. The popular impression that the Southern Californian is self-selected for certain personality traits merits careful investigation.

In any case, the basic condition for the voluntary region appears to be met: the replacement of social regions determined by the traditional factors of circumstance of birth and social heredity by self-selected groups of like-minded, mobile, atomistic individuals. The cultural personality of Southern California is decreasingly defined by its nineteenth century infancy; instead, there is constant redefinition with the arrival of more strangers and the formulation of new attitudes and cultural entities, as some still obscure new processes become ascendant. In the case of peninsular Florida, there is virtually no nineteenth-century residue to be expunged. The great influx of settlers—many, of course, from the South, but a large fraction from the non-South (or the Caribbean)— have been continuously improvising a fresh cultural milieu. We still have no clear picture of the emergent regional culture and subcultures of an area with no genuine *raison d'être* except the presence of people; but there is little doubt that the many hundreds of thousands of persons pursuing the varied amenities of Florida and their own self-realization are building something far different from the culture areas of colonial America or premodern Europe.

At the micro level, we can assume a large, diversified array of voluntary regions. Here we shall briefly note only a few categories: those that are conspicuously visible and also adhere, at least partially, to the conventional rules of spatial contiguity and concentration. A number of possibilities are omitted because their existence is still too speculative and their unusual spatial attributes too difficult to specify with the limited means at our command. Let us proceed from traditional examples to the relatively avant-garde.

MILITARY SUBREGIONS. There is, of course, nothing particularly newsworthy in the existence of a professional military population with its own way of life and discrete spatial enclaves. We can readily find analogies in many other countries, some much less advanced in development than the United States. We can also assume a universal tendency for the military subpopulation to be self-selected for certain attitudes or personality traits, over and above purely mercenary considerations. But what tends to be special in the United States is the sheer size, solidity, territorial extent, population size, and local social impact of the larger military establishments. Thus a special subculture suffuses and tends to dominate several sizeable metropolitan areas, for example, San Diego, Colorado Springs, Columbus, Georgia, Norfolk, Annapolis, and Pensacola. This is also true for paramilitary research and development centers at Los Alamos, Oak Ridge, Huntsville, or Cape Kennedy. All these places, along with many a lesser locality, actually constitute one large, discontinuous region, or network of points, whose various interchangeable components are in constant, intimate connection with each other, more so perhaps than with other spatial systems.

EDUCATIONAL SUBREGIONS. The close connection of many widely scattered points is even more obvious when one observes our hundreds of colleges and universities with their faculties, students, and hangers-on, including a growing cohort of the retired. As elsewhere, economic factors are present, but the decision to live or work in a college community, or in which specific one, is increasingly subsidiary to matters of taste and personality. We can also identify analogies abroad and scattered local examples in our own earlier history. But, again, the massiveness of this population, involving some millions of students and faculty members, their rapid growth since 1945, and intense social and cultural development of our many college towns and college ghettos within larger metropolises has meant a new, truly unprecedented order of cultural subregions. Not every college automatically generates its own microregion: many are too weak or thoroughly overpowered by the larger society, or are frequented by commuting student bodies and faculties. But if many of our burgeoning institutions of higher learning have acquired and staunchly cling to their own idiosyncratic personalities, the similarities among these places and their people are far closer than those with local traditional regions or with other spatially discontinuous, national subcultures. Like most of the remainder of this discussion, these remarks are impressionistic. The social and cultural geography of college populations is still almost totally *terra incognita.*

THE PLEASURING PLACES. The richest category of voluntary regions includes all those many places and people who, in the wider sense of the terms, are in quest of pleasure, surcease, or self-improvement. Assuredly, no other nation or epoch has spawned such a number or variety of amenity-oriented communities, ephemeral or not. The following subcategories do not begin to exhaust the possibilities. Nor are they mutually exclusive; many localities fall into two or more classes.

In a few years, we can expect to have virtually every physically attractive *amphibious region* thoroughly exploited and seasonally saturated with patrons. As it is, there are few suitable, undeveloped, privately owned tracts remaining along the Atlantic, Pacific, or Great Lakes coasts, and the Gulf Coast is also filling up rapidly. The same story can be told of the glaciated lake country of the Northeast and the lakes and artificial reservoirs in other parts of the country. During a warm season that varies from scarcely two months in the far north to the entire year in the far south, uncounted millions of Americans gravitate to the waterside to swim, fish, boat, surf, beach-comb, sun-bathe, cavort in the sand, or just stare at seascape and shore. Each of these communities, generally stretched out along a beach and only a few hundred yards in width, tends to have its own social characteristics and to foster a peculiar subculture. Certainly as a group the amphibious communities claim a personality all their own; and many, although only seasonally alive, are so vigorous as to dominate the entire annual cycle of their localities.

Many of the *heliotropic* (sun-seeking) *regions* also fall into this category, but others, especially in the relatively warm, cloudless Southwest, are high and dry. In turn, a very large fraction of the *retirement subregions*, notably those in Florida, Arizona, and California, are either heliotropic or amphibious. But a considerable number of these are being deliberately engineered, mainly for commercial gain, in other parts of the country. Such spatial segregation of the aged (on a scale beyond the conventional old people's home) is a novel phenomenon, one still rather neglected by geographer and sociologist.

All those accessible *montane regions* with adequate slopes and natural (or artificial) snow potential within range of large metropolitan populations eager for skiing and other winter sports have undergone rapid development in recent years, especially in New England, New York, and the Sierras. Parallel development also proceeds in sections of Colorado and Idaho, where skiers arrive mainly by jet. Indeed the elaboration of the ski subculture, along with various summer recreational activities, has been so intense in much of Vermont and nearby states that a genuine transformation of the cultural character of that area may be in progress. The socioeconomic impact of the pleasure-seekers here, as in other pleasuring places, has been quite profound.

The one subcategory of amenity-oriented subregion that seems well-rooted in European antecedents, aside from the dwindling medicinal spas,[33] are the American *equine regions*. They are, or at least have been, markedly and affectedly British in style. There are now a goodly number of such areas, where the breeding and rearing of horses, riding, and hunting literally constitute a whole way of life. The strongest development appears in southern Megalopolis, especially in parts of Piedmont Maryland and Virginia that are suburban to Washington and Baltimore; the Bluegrass; and a tract in northern Florida.

The subcategory that in lieu of a better label is termed the *forbidden fruitlands* is not peculiar to North America; but again there is no development elsewhere comparable in scale. These are the areas where social and economic existence is based upon activities classed as illegal in most states or municipalities—gambling, prostitution, nudism, unrestricted

[33] David Lowenthal, "Tourists and Thermalists," *Geographical Review*, 52, No. 1 (1962), 124–27.

liquor consumption, open homosexual behavior, and certain sports. Beyond any question, this is how Reno, Las Vegas, the Lake Tahoe resorts, and the other urban centers of Nevada must be regarded, or the string of international twin cities along the United States-Mexican boundary, and a variety of satellite towns along several of our interstate lines. Again, the disreputable activities may be combined with more wholesome ones in various hunting and fishing or seaside areas. The transient subpopulations who frequent such areas and the service groups who cater to them act to create a special cultural ambience that has yet to be analyzed geographically.

Still within the realm of pleasure, there are many other temporary aggregations of persons of common taste or purpose who may manage to impose more than a fleeting impact upon a considerable locality: the rallies of cyclists or hot-rodders, "wilderness" camping, rock festivals, summer music camps, revival meetings, and the like.

LATTER-DAY BOHEMIAS AND UTOPIAS. The last pair of spatially discrete sets of voluntary regions are unquestionably North American in character, if not in ultimate origin, then certainly in their current full-bodied expression. The first of these is so very recent that scarcely any academic notice has been taken of them (and they still lack a suitable generic term). The "New Bohemias" are those quite distinctive neighborhoods toward the center of our larger cities that have attracted such urban folk as actively relish diversity, unconventional behavior, and a variety of exotic stimuli. They are complex in ethnic, racial, class, and age structure, and (with the possible exceptions of the prototypical Greenwich Village in New York and North Beach in San Francisco) are products of the 1950's and 1960's. They are still emerging, spontaneously or by design. Perhaps the best way to define them is to cite examples. In addition to Greenwich Village and North Beach, there is Chicago's Old Town, the Beacon Hill-Back Bay section of Boston, Atlanta's Peachtree Street, N.E., the French Quarter of New Orleans, Rittenhouse Square in Philadelphia, Georgetown and Dupont Circle in Washington, and Cincinnati's Mount Adams. In many places, such as Berkeley, Seattle, Minneapolis, or Toronto, the special university neighborhood coincides perfectly with the New Bohemia; but in most, the two sections lead quite separate existences. One can anticipate a further proliferation of this novel form of urban life not only in the United States but in other highly advanced nations in the near future.

During the past decade, a great number of *communal colonies* have sprung up in virtually all parts of the country. Examples abound in urban as well as rural settings; they vary considerably in size, mode of organization, style, purpose, and type of membership. By their very nature (many go to some pains to escape detection), they are not readily susceptible to synoptic analysis. Perhaps the only attributes they share are communality of property, a rejection of the regnant values and rules of the greater world, and an adherence to the principles of simple living and decent poverty. What is not clear is whether these latter-day communes are the natural progeny of the numerous utopian experiments, largely religious or political in character, that flourished in large numbers, though briefly, during the past century, or whether they have sprouted spontaneously out of the inner core of American culture. In any event, it remains to be seen what, if anything, they herald of the future shape of the American cultural landscape.

The foregoing sketch is confined to what is known or reasonably assumed to exist and to those entities within the family of voluntary regions that can be readily plotted on the conventional two-dimensional map. Possibly they are only the leading edge, the most visible, mappable forerunners of a large new galaxy of cultural associations that observe an as yet uncharted code of spatial rules. We must be alert to the emergence of unexpected modes of cultural interaction within the American realm. Assuredly we are entering, or are well into, a period of shattering change in the social organization of highly advanced communities, and in the rapid unfolding of technologies and mental worlds still scarcely dreamed of. It is axiomatic that spatial structures and processes are interdependent with the basic social and perceptual attributes of a people. It would be strange indeed if we did not soon find ourselves stumbling across a threshold into radically unfamiliar ways of locating our cultural and other experiences on this fragile earth-shell as we grope our way so painfully toward an undisclosed destination.

Postlude

Anyone who has flipped to the back of this volume to find an answer to the question posed in its first line will be doubly disappointed. There is no valid answer, and there is no real conclusion—only a pause. These closing lines are offered in lieu of a proper introduction: an author is often not fully aware of what he wishes to say until his manuscript tells him; and the reader is seldom primed to appreciate an introduction until he has traversed the main body of a work.

This has been an attempt to organize an important, but previously unorganized subject, the cultural geography of the United States. As the first such effort, it is subject to many hazards. Not the least of these, all too obviously, is the writer's frailties, including major zones of ignorance and many instances of personal bias. No apology is needed, however, for one basic decision, namely to eschew the discursive, encyclopedic recounting of the cultural features of the nation. The materials for such a panoramic treatment, or at least many aspects thereof, are reasonably abundant. The reader in need of factual nourishment is earnestly directed to the selected and annotated references that follow; indeed this list is an integral part of the volume and should be foraged with care.

The design chosen for this essay seeks to interweave the theoretical with the particular so as to shed light on both. This is done in the conviction that serious attempts to combine theory with the regional method have been all too rare or timid in modern geography, and that much is to be gained thereby. But in view of how little is truly known by either anthropologist or geographer about the basic processes that mold the cultures of large nations or ethnic groups, or their constituent regions, this volume is interrogatory in tone. I have settled on four basic questions: Culturally speaking, who are we Americans? What are the sources of our identity? What processes, operating how, when, and where, have molded our cultural patterns? And what spatial configurations have emerged from such activity? In experimental fashion, I have offered one tentative way of seeking answers, but without any illusions that this scheme is either final or impregnable to challenge. Confining an unruly reality within a simple conceptual harness is an impossible chore. But I do hope that this perspective on the American cultural landscape has at least the merit of provoking better, clearer ones. In any event, it seems to accommodate a maximum of fact with minimal mutilation; it illuminates many murky items, and it suggests many new questions. I anticipate better schemata in the near future, and await them with interest. Lasting consensus (or tolerance) is sought from readers, critics, and successors on only four points:

1. The cultural patterns of the United States are exceedingly interesting in themselves. Infatuation with any and all of their details is a commendable form of madness.
2. Such patterns are important not only to Americans but to the contemporary world in general. It is urgent to get at their fundamental structure in the cause of planetary social and physical survival and the fostering of civilized existence.
3. Both the spatial and temporal texture of cultural events must be apprehended before the inner logic or laws of cultural transactions can be spelled out. Where, when, and how form a sacred threesome, triune yet unitary.
4. We live in a period of rapid structural change in terms of ongoing cultural process in the United States and of penetrating the most fundamental questions concerning human thought and behavior, as they are constrained within the cultural matrix. We can also anticipate a period of almost revolutionary insights that will vastly enrich our understanding of what we have been, who we are now, the most likely forms for future cultural dealings among people, places, and things, and even how they can be manipulated so as to give ourselves a fighting chance for survival and salvation.

PART 3 afterthoughts

CHAPTER FIVE

America in Flux

Introduction

The business of this final section, or postscript, is twofold in character: to describe, explain, and interpret, as best one can in the throes of swift and bewildering change, what has been happening to the cultural patterns of our country over the past twenty years or so; and, then, to note some of the things the cultural geographer has learned about these new developments as well as the unfinished agenda of earlier times. In short, what can we say about the cultural geography of the United States from the latter-day perch of the 1990s that could not be articulated in 1973?

Retrospection has certain virtues, of course. What may have been obscure a generation ago becomes clearer in hindsight—not to mention all those items I should have noted back then but was too obtuse to discern. Moreover, the quantity and, I dare say, quality of relevant research has grown markedly in recent years as geographers and kindred souls have begun to pose a wider range of worthwhile questions and to explore them in greater depth. Although the temptation to chronicle the recent achievements of my colleagues methodically and at some length is real enough, I shall refrain and devote only a single paragraph to such intramural matters. Instead, the plan is to concentrate topic by topic on those areas where cultural and social evolution has been significant and interesting, while pointing out in passing such related geographic and other scholarship as may be worth noting.

The sequence of items that follows is as logical a one as I can devise, but, as the discerning reader will discover, the constraints of language, being the linear medium that it is, prevent one from doing justice to the multidimensional interconnections of a tangled web of phenomena. In order to squeeze maximum meaning from the material at hand, I must switch among global, national, and regional/local frames of reference at certain points. Indeed, there is no sensible alternative to situating present-day American cultural behavior in the context of a highly interactive world system. Whatever simulation of self-containment may have been workable in generations past is no longer valid. In a parallel strategy, one that further distances this edition from the earlier version of the book, there is greater reference to nongeographic and noncultural realms of reality. Thus I wish to set the spatial and temporal aspects of our American cultural system within the matrix of larger forces, of the actualities of an ever-more interdependent world.

Cultural geographers in the United States and elsewhere may be most reluctant to admit it, but the final third of the twentieth century is proving to be our golden age. Although we (and the sibling guild of historical geographers) may be suffering from a certain Cinderella Complex as some trendier geographic subdisciplines garner more disciples, lucre, and general glory, the recent accomplishments of the cultural-geographic community have been substantial. The number of practitioners may have expanded only modestly, but the volume, scope, and sophistication of their output can justify some congratulation. New associations—formal and ad hoc—have crystallized, as have new journals, and a generation of atlases, monographs, and symposial volumes well beyond precedent. We have begun flashing light into hitherto shadowy corners of the cultural cosmos. In part, this is the result of circumambient social, political, and other broader developments or the gentle pressure of neighboring disciplines. But we can also ascribe such widening arenas of curiosity to a general maturation of thought and method and the automatic internal dynamics of intellectual evolution. A few of the new paths along which promising starts are visible include the geography of sport, foodways, and music; the intersections between literary fiction and place; and exploration of the hidden cultural messages of maps. Even more heartening is the flourishing condition of the study of the built landscape, including, belatedly, such items as commercial structures, cemeteries, monuments, and signs.

The Ambient Scene

There is no way to avoid glancing at the overall conditions of contemporary American life before narrowing our attention to specifically cultural matters. Such an exercise obliges me to contrast our situation today with an earlier scene. And, for the most part, it is a rather sobering experience. By pure happenstance, the writing of the first four chapters of this volume took place during a period (1968–1972) of exceptional social and political turmoil and national soul-searching. If it was a troubled period, it was also one that seemed to hold the possibility, disturbing to many, welcome to others, of considerable, possibly revolutionary, alteration in collective attitudes and behavior. It is fair to say that many of these expectations have been frustrated. But, in any event, it was feasible then, as it had been in even the darkest moments of the Great Depression of the 1930s, to assume that the congenital American faith in progress, in unending social and material betterment, was not a total delusion. How many of us today remain honestly committed to that credo?

If it was once possible to write in good faith, as I did in the initial version of this volume, that "Quite clearly the United States is the most powerful nation of the contemporary world in both economic and military terms. This nation also retains much of its traditional strength as a moral force, and has had few recent rivals as a generator of science and technology . . ." (p. 5), today, sadly enough, any such statement must be amended.[1] Real personal income, perhaps as meaningful a measure as any of general well-being, reached its apparent all-time peak in or about 1973, and since then it has stagnated or, more likely, has declined appreciably. That is not to say that certain sections of the country have not been flourishing in recent years or that a goodly number of individuals are not faring better now than formerly, for such is clearly the truth.

The larger truth, however, is that, in contrast to previous generations of American parents, most of us *cannot* expect our children to enjoy greater material satisfaction or to attain higher social or economic status or self-realization then we have known.[2] A full explanation of this turn of events, including the decline of the dollar and the dubious distinction of America becoming the world's greatest debtor nation (and a concomitant restructuring of the economy and labor force), is well beyond the scope of this discussion. But undoubtedly one of the major factors is an increasing integration into, and dependence upon, an ever-more interdependent world capitalist system. Consequently, many decisions affecting the livelihood and lifestyles of Americans are now being made by foreign governments (especially those with lighter military burdens or something other than our political arthritis) or by business firms with no special allegiance to any country—just as in the past (and to some extent today) the workings of American businesses have impinged upon the fortunes of hundreds of millions elsewhere. The massive incursion into American life of foreign capital, factories, and other operations and their products is related to the equally visible transnationalization, or sharing, of cultures (something we shall note more particularly at a later point), and has not escaped general notice. Nevertheless, the most basic of the newer facts of American life—the prospect of prolonged austerity or worse—has not yet fully sunk in. When it does, the consequences for our cultural geography at both the national and local levels should be well worth studying.

One result of a creeping economic malaise, already all too apparent to the observer, is the increasing shabbiness or actual deterioration of our public facilities and the common spaces we are obliged to share. Once

---

[1] The "American Century," trumpeted so lustily by Henry Luce in the 1940s, lasted approximately 25 years.

[2] No need to burden the reader with the statistical evidence supporting this argument. It is ample enough. More telling perhaps are such offhand observations as the following item in a dispatch by John Lukacs from a hyperprosperous Germany: "Has it ever happened, in this century, that an American traveling abroad found that he was poorer than were the natives of his class? Perhaps less elegant, less sophisticated, less polished, less worldly. But poorer? This in not an ephemeral phenomenon. It is replete with a meaning . . ." *Harper's*, 282, No. 1690 (1991), 67.

again, of course, there are the exceptions, those oases of affluence and privilege where individual wealth or well-heeled private governments hold sway; and not all places are as poorly off or as wretchedly maintained as the notorious decaying cores of certain metropolises within the so-called Rust Belt. The litany of crises is becoming familiar and by no means confined to the more distressed sections of the nation. In physical, that is, infrastructural, terms, it includes crumbling roads, bridges, and waste disposal systems; littered roadsides and derelict properties; substandard park and school building maintenance; worsening air quality in many localities; problems with water supply and its potability; inadequate housing stock for the less fortunate; and obsolescent, fearfully overcrowded prisons. The contrast with much of Western Europe and certain other favored lands is becoming embarrassing. However, the outlook need not be entirely dismal. Discontent with the inadequacies of our institutions may ultimately generate corrective measures, just as realization of the finiteness and fragility of our ecosystem has recently compelled us to begin reversing the ways we behave, singly and collectively.

The various ills noted above are not unique to the United States since we can recognize them as occurring with greater or lesser severity in other regions of the First and Second Worlds. Thus somewhat similar dilemmas vex the societies of Great Britain, Ireland, Argentina, Poland, and the Soviet Union, while nearly all the nations of the Third and Fourth Worlds are sorely afflicted in ways that are all too visible. Evidently some sort of systemic crisis is overtaking the entire human enterprise, however unevenly it is expressed spatially. But there is one extraordinary attribute that sets America apart from all other lands.

MILITARISM AND VIOLENCE.   It pains me greatly to bring up the inescapable: the fact that the United States has become essentially a garrison state, that the militarization of the economy (with the predictable fallout in our collective attitudes) and thus the institutionalization of violence has become an accepted, defining condition. Whatever our other failings, the United States has demonstrated to everyone's satisfaction that it is indeed the world's strongest military power as of the 1990s. Whether fiscal and political factors will prevent such preeminence from being sustained indefinitely is beyond anyone's powers of prediction. But history does offer some solace: the example of radical turnarounds in the past. For the first 165 years of national existence, America's standing military establishment was diminutive by conventional Great Power standards and almost negligible in its impact on the larger economy and society or political arena.[3] And then we have the transformation of the ferocious Swedes of the seventeenth century, the bellicose Germans and Japanese of the pre-1945 era, or the nettlesome Swiss of yore into the regular pussycats they are today. But we also have the metamorphosis of

---

[3] The Grand Army of the Republic, formed by Union veterans shortly after the Civil War, was a political force for some years, but a transient one, fading away as it did after the turn of the century. For the recent militarization of the economy, see Ann Markusen et al., *The Rise of the Gunbelt: The Military Remapping of Industrial America* (New York: Oxford University Press, 1991).

the meek, nonviolent Jew of venerable tradition into the formidable Is-
raeli warrior of today.

A militarized America is a post-1945 phenomenon. Although various
victorious generals and admirals may have become national heroes and
ascended on occasion to the presidency or other lofty positions, and although the martial spirit did flare up sporadically, the early United States
was basically antimilitary in mood, harboring deep hostility toward standing armies and the officer caste.[4] It was not until 1940 and the waging of
World War II, the Cold War, and a succession of bloody foreign adventures, of which Korea, Vietnam and the Persian Gulf were on a massive
scale, the maintenance of an enormous military presence at home and
abroad, along with a vast and costly R & D (research and development)
industry beholden to the armed forces, that the Military-Industrial Complex became a seemingly eternal fixture on the American scene.

There are at least two good reasons why cultural and other geographers should pay attention to the facts of American militarization. Most
immediately, we have all those still poorly studied landscape signatures
of the military: the many army, navy, and air force installations; recruiting
stations and posters; military cemeteries (an American innovation); battlefield parks[5]; veterans clubhouses; monuments and memorials with a
military message; or the epidemic flare-up of flags and yellow ribbons in
all manner of places in early 1991. The regionalization and other spatial/
social correlates of all such items merit looking into. But beyond whatever analytical value such tangible evidence may present, there is a
deeper significance in the latter-day role of the military in American life,
something that lies beneath the surface. An inescapable prevalence in
contemporary America, a more than passive acceptance—indeed, quite
often the celebration of the military ethos (as the only usable device for
activating a sense of national community?)—would scarcely be possible if
a certain predisposition had not been present. And such a predisposition,
I suspect, is linked with one of those facets of the national character
barely hinted at in a previous chapter: a reliance on violence as an effective and normal instrument for social action.

Obviously Americans, past or present, have never enjoyed a monopoly over the use of violence, or the tolerance thereof. Virtually all other
societies have had their outbursts, and indeed the very existence or credi-
bility of any state as we know it is predicated on its capacity for applying
physical force within or beyond its borders. Furthermore, the chronicles
of other European settlers and operations overseas are replete with violent acts. But the circumstances surrounding the settling of America, its
unusually early attainment of political autonomy, and the ways in which
its cultural system was formed all fostered an addiction to violence that
seems qualitatively different from the situation in other settler countries

---

[4] Marcus Cunliffe has amply documented this attitude in *Soldiers and Civilians: The
Martial Spirit in America, 1775–1865* (Boston: Little, Brown, 1968). Especially symptomatic
of the general mood was the persistent antipathy toward the Order of Cincinnatus, an organization of Revolutionary War officers and their male descendants.

[5] Reuben Rainey, "The Meaning of War: Reflections on Battlefield Preservation," in
Richard L. Austin et al., eds. *Yearbook of Landscape Architecture* (New York: Van Nostrand
Reinhold, 1983), pp. 69–89, is a thoughtful initial effort in understanding the cultural impli-
cations of battlefield parks.

(Israel perhaps excepted) and certainly from that in older Old World national communities. The safeguards of well-entrenched social constraints in the latter and the moderating influence of long-distance governance in the former, along with the bestowal of venerable institutions, meant the relegation of violence to a subsidiary role in the course of everyday affairs.[6]

If an American value system, a rambunctiousness that condones, and occasionally even glorifies, violence and vigilantism may help account for the recent rise of militarism, it also finds expression in many other ways that should kindle the curiosity of the cultural geographer. We have already seen that there is an interesting geography of homicide in America, but without noting a national rate wildly in excess of that in all foreign competitors. There is certainly a cultural component in those other forms of violent crime to which Americans can resort, and intriguing regional and local geographies thereof remain to be investigated.[7] The public reaction to crime is a resort to retaliation rather than cure or prevention, a punitive strategy that has produced a rate of incarceration seldom equalled elsewhere. Furthermore, the United States is one of only two countries, South Africa being the other, in which capital punishment is still legally practiced (however murderous certain other regimes may be extralegally). The incidence of private firearms (and their use in criminal pursuits and personal altercations) is also off the international scale. And where else would a Gun Lobby have such political clout? It follows that the prevalence of hunting may be greater in the United States, in both absolute and per capita terms, than anywhere else. Resort to armed struggle is an article of faith among the extremists of the far right and the occasional group at the other end of the political spectrum, an uncommon situation in most other countries at comparable levels of socioeconomic attainment (Northern Ireland being the notorious exception). The geographer can find tempting opportunities for research in all such matters.

The less direct expressions of violence may be especially symptomatic of what lies at the core of our collective being. Thus the theme of physical conflict has been popular in pulp fiction and other types of printed excitement for the masses ever since the formation of a national market for such reading matter. More recently, however, the incidence of violence and bloodshed in films, television programs, video games, and comic books has reached levels that many regard as excessive and psychologically harmful. Some critics have been particularly disturbed by the

[6] The comparison with Canada, our next-door sibling and a country so similar, yet so profoundly different in its social and cultural axioms, is peculiarly instructive. Some of the reasons for and expressions of these differences are discussed in Pierre Berton, *Why We Act Like Canadians: A Personal Exploration of Our National Character* (Toronto: McClelland and Stewart, 1982); Michael Goldberg and John Mercer, *Continentalism Challenged: The Myth of the North American City* (Vancouver: University of British Columbia, 1985); and Wilbur Zelinsky, "A Sidelong Glance at Canadian Nationalism and Its Symbols," *North American Culture*, 4, No. 1 (1988), 2–27.

[7] The rich potentialities of such a line of research are apparent in two pioneering volumes: Keith D. Harries, *The Geography of Crime and Justice* (New York: McGraw-Hill, 1974) and Keith D. Harries and Stanley D. Brunn, *The Geography of Laws and Justice: Spatial Perspectives on the Criminal Justice System* (New York: Praeger, 1978).

Saturday morning slaughterhouse on television that diverts so many millions of youngsters. Finally, is it too far-fetched to characterize the rising tide of litigation and all those courtroom battles in America (and the accompanying proliferation of lawyers) as a sublimated form of violence?

NATIONALISM. If militarism and violence are American phenomena too pervasive to ignore, we must also come to terms with a related social fact of even greater potency: Americanism, the term for our local brand of nationalism, or statism. That shared emotion and set of practices we call nationalism has become the dominant faith of most populations of our twentieth-century world. In many instances, the American case among them, nationalism has replaced or expropriated the traditional supernatural religions to become the strongest available adhesive to bind large communities together and bend them to some sort of common purpose.[8]

I have come to believe that we cannot attain an adequate understanding of American history, geography, culture, and much else unless we probe the nature and workings of our nationalistic sentiments. And since nationalism is so thoroughly symbolic in nature, I have examined and interpreted all the accessible symbolic evidence in the form of ideas, ideals, behavior, and objects; then, plying the craft of the historical geographer, I have plotted their spatial and temporal ranges.[9]

Before summarizing the results of this investigation, results that were rather surprising, it is essential to specify the two subcategories I call *statism* and *nationalism* and their strict meaning since they are generally subsumed under the generic term *nationalism*. One must realize, to begin with, that *state* and *nation* are not at all synonymous. The former is fairly easy to define. A state is a political apparatus that claims or exerts absolute sovereignty over a given territory and its inhabitants, and as such it has existed in various areas for some millennia. Defining the *nation* is a much trickier business, but in essence we are dealing with a sizable collection of individuals, with or without aspirations to statehood, who feel united in having a unique and cherished social and cultural identity and set of traditions.[10] (According to this definition, the terms *nation* and *ethnie*, or *ethnic group*, are virtually identical in meaning.) In its purest sense, then, nationalism means intense devotion to the nation, that real or supposed community which regards its unique set of shared attributes and memories as something more precious than life itself.[11] When nation

---

[8] Although the two phenomena obviously overlap, militarism and nationalism are by no means identical. One can be militaristic without being devoted to a particular national ethos, and a peace-loving nationalist is quite imaginable.

[9] Wilbur Zelinsky, *Nation into State: The Shifting Symbolic Foundations of American Nationalism* (Chapel Hill: University of North Carolina Press, 1988).

[10] Ethnicity, as we know it and think about it today, is essentially a product of modern times. Although there may have been certain dim stirrings of such cultural self-awareness many centuries ago, it could not develop fully until communities had crossed a critical threshold of socioeconomic evolution.

[11] I have avoided using the term *patriotism* here for the simple reason that it is by no means synonymous with nationalism. The former means devotion to one's locality rather than to any larger territorial or political entity, as illustrated by the quite different connotations of *pays* and *patrie* in the French language.

and state are effectively combined and merged within a given territory, we achieve the modern *nation-state*, which is so beautifully embodied in such examples as the United States, Japan, Israel, Poland, or Ireland, and is an entity that originated no earlier than two hundred years ago. The fully successful operation of the nation-state demands, and receives, the wholehearted loyalty of its citizens—indeed, an incandescent level of adoration that we can designate as *statism* or *statefulness*. The acolytes thereof perceive the state (if not necessarily the government) as the ultimate social reality, as the repository of all that is precious and noble in life, and are quite prepared to lay down their lives for it when such an ultimate sacrifice is necessary.

Despite their recent merger in so many instances, it is essential to realize that nation and state, as well as nationalism and statism, are inherently distinct in character and origin. And the basic difference lies in the level and direction of the flow of grace and power. While the nation may be a real or imagined freemasonry of brothers and sisters, a more or less egalitarian confederacy, conjoined through blood, soil, "the mystic chords of memory," or some other web of sentiment, the state floats far above the grasp or control of the common herd: stern and austere, though nurturing; majesty rather than fraternity; compulsion, or surrender, replacing mutuality.

The preceding observations apply reasonably well to almost any part of our modern world. But what do we find by narrowing our attention to the American scene? As it turns out, the result is a periodization over time of the nature and intensity of nationalistic and statist behavior. Whether such a progression is observable in other nation-states is a question only further research can answer. In any event, the temporal patterning—that is, the appearance, rise, climax, and eventual decline and possible disappearance of some seventy-odd cultural phenomena of a nationalistic import that I have documented—fits rather convincingly into three general phases or periods. Each of these epochs has its peculiar defining characteristics, but no specific date at which it begins or ends. Transitions are too gradual and the overlapping of some attributes large enough to make firm dating impractical.

The first phase lasted from the 1760s or 1770s through the 1840s, those early decades of imminent, then actual, independence when a pure, pristine nationalism was exceptionally strong and as yet unbesmirched by statism. It surpassed in vigor all competing emotional forces except the enthusiastic Christianity with which it was entwined in a mutually nourishing symbiosis. The fragile, emergent state was tolerated, being respected to the degree that it embodied the lofty national principles. A faith in the unique virtues and transcendent mission of the American Republic, deeply internalized, filled the hearts and minds of most men and women. Americans expressed their nationhood (but not statefulness!) through, inter alia, a distinctive vernacular architecture; hero-worship; singing appropriate songs; adopting nationalistic names for persons, places, and other objects; observing the great national holidays; and revelling in the paintings, oratory, and printed matter that extolled their remarkable new country. If a date must be set for the apogee of this emotional surge, let it be 1824–1826. We had then the near-coincidence of Lafayette's return visit (1824–1825), which set off an orgy of celebration unsurpassed before or since, and the awe-inspiring simultaneity of the deaths of John Adams and Thomas Jefferson on July 4, 1826, the fiftieth anniversary of the Declaration of Independence.

With the fading away of the Revolutionary generation and its living memories of the transforming early struggles and providential blessings, not to mention the arrival of sectional strife, massive immigration, industrialization, and wrenching economic and social innovations, American nationalism began to experience its second phase by the eve of the Civil War. Nationalism, now blended with a nascent statism, retained much, but not all, of its early élan. The heroes, deeds, and ideals upon which was grounded the nationalist creed of the initial generations of Americans had slipped beyond immediate reach, receding into a remoter emotional space. Thereupon the nationalistic monument entered the scene as a device both necessary and effective for jogging collective memory. The eagle, originally an item bursting with libertarian messages, mutated into a content-free emblem, one that, like the flag, simply declares identification with the state, and little more. Similarly, the neoclassical style, once so laden with ideological baggage, dwindled into one of several architectural clichés, but was the one style most serviceable, like flag and eagle, for connecting a building with a pervasive federal regime, with centralized authority, rather than any special principle. Such innovations as the world's fair, historic preservation, Uncle Sam (usurping Miss Liberty), state funerals, national parks, and an imposing array of federal structures were redefining the American *mentalité*, along with a presidency that was undergoing sanctification. The 1890s may have been the decade when this second phase blossomed most luxuriantly.

The third, and current, period of American nationalism—its emphatically statist phase—may have germinated as early as the 1880s, but did not ripen fully before the late twentieth century. Insofar as it harkens back to any past, it has been a synthetic, nostalgic nationalism no longer rooted in the historic actualities that bred its predecessors, and one with little or no awareness of the contentious issues that made early America such an unusual, often subversive place in the eyes of the world. To the degree that any traces of the primordial nationalism still linger on, it is a phenomenon no longer at odds with American statism, a conflict that could quite possibly have been simmering below the surface in earlier times. Instead, the two strains of sentiment now cohabit snugly. And thus the United States has become a full-fledged nation-state. The symbols, rites, and other nationalistic performances in vogue today are essentially nonideological and nonethnic in character, simply proclaiming the power and glory of Uncle Sam. Quite clearly, too, the quotidian emotional temperature of American nationalism is currently much cooler than in its heyday (and such is also the case for statefulness except in moments of crisis), if for no other reason than that of the many new competing demands on the time and psychic energies of a fully modernized, complex society.

Like the active practice of most traditional religions, the observances

152   *America in Flux*

of nationalism have become a sometime, rather tepid thing. For Americans coping with a bewildering world of shifting values, the essence and uniqueness of America are now items for casual viewing at museums and costumed reenactments; spiritual refuge is available in the arms (literally and figuratively) of the state. Showing the flag, especially during periods of international altercations, or at Olympic Games and the like, will have to do as the antidote to doubts about purpose or identity. On a daily basis, we pay homage to the presidency and quadrennially experience the orgy of an inauguration. In the contemporary world, with its strange mixture of sophistication, cynicism, and wishful dreaming about the past, the traditional nationalistic monument is obsolete; and when something absolutely must be built, the approved design is fashionably ambiguous. What is more generally useful, however, is the artfully framed, illusionist historical museum in all its many forms and, for the vernacular structure, a Colonial Revival treatment, the mythopoeic reshaping of the past as it should have been.

Because the evidence so far gathered and studied is fragmentary, we must be more cautious in speculating about the historical geography of nationalism in the United States than about its history. Nevertheless, such data as we do have all point toward a single hypothesis: that nationalism bloomed earliest and most vigorously in the New England and Middle Atlantic states, then subsequently grew lustily in those sections of the Middle West settled from this northern segment of the Atlantic Seaboard, while it lagged far behind in the South (which, arguably, was busy contriving its own alternative version of nationalism). The present-day (and presumably earlier) incidence of flag and eagle display and holiday observances by region supports this notion, and a complete survey of nationalistic monuments is likely to provide further corroboration.[12] The historical geography of nationalistic place-names and classical town names admits no other interpretation, while studies of three phenomena with decided psychological linkages to the specifically American brand of nationalism—political cultures, utopian communities, and old-line fraternal orders—all indicate sharp North/South disparities in keeping with the spatiotemporal scenario sketched above.[13]

Although the specifics of this story may be uniquely American, other nation-states have also undergone somewhat analogous transformations over time, whether the entity in question began as nation or as state or even, in some instances, both nation and state appeared at roughly the

[12] For an introduction to the geography and history of flag and eagle on the American scene, see Wilbur Zelinsky, "O Say, Can You See? Nationalistic Emblems in the American Landscape," *Winterthur Portfolio*, 19, No. 4 (1984), 77–86.
[13] Wilbur Zelinsky, "Classical Town Names in the United States: The Historical Geography of an American Idea," *Geographical Review*, 57, No. 4 (1967), 463–95 and "Nationalism in the American Place-Name Cover," *Names*, 30, No. 1 (1983), 1–28; Daniel J. Elazar, *American Federalism: A View from the States*, 2nd ed. (New York: Crowell, 1972); Philip W. Porter and Fred E. Lukermann, "The Geography of Utopias," in David Lowenthal and Martyn J. Bowden, eds., *Geographies of the Mind* (New York: Oxford University Press, 1975), pp. 197–233; Richard H. Schein, *A Geographical and Historical Account of the American Benevolent Fraternal Order*, Master's thesis (University Park: The Pennsylvania State University, 1983).

same time. The American case is unusual in the relative transparency of the process, one dusted only lightly with the detritus of ancient history or the complexities of antecedent local cultures and politics. In any event, the final, or contemporary, outcome is a statism closely resembling that of other relatively stable, mature countries, however special the indigenous symbolic and other details may be. It is a fact as fundamental and general to the understanding of present-day cultural geographies as latitude, longitude, geology, landforms, soil qualities, or population structure are to the study of physical or economic geography. But my compatriots may take some pride in the fact that, as the "first new nation," the United States was obliged to blaze a trail, presenting something of a model for other emerging nation-states to emulate, as it invented, among other such exportable notions, the national hero, national holidays, the very first synthetic capital city, flag idolatry, veterans associations, battlefield and national parks, and those powerfully symbolic military cemeteries.

EFFECTS OF TECHNOLOGICAL CHANGE IN COMMUNICATIONS AND TRANSPORT. Culture and technology have always interacted with each other, in ways both gross and subtle, from prehistoric times onward; but in no era has this truism been more blatantly alive, more pertinent, than at present. An ongoing revolution in communications is the most vivid illustration of the principle. Much remains to be learned as to just how the advent of universally available postal service, including Rural Free Delivery, electric telegraphy, the telephone, photography, typewriter, and nationwide distribution of books, periodicals, and catalogues—along with all those standardized, mass-produced artifacts that also convey much information—may have altered cultural patterns in nineteenth-century America; but certain it is that the relationships have been both complex and major as what might be termed the opening phase of the modern Information Age took shape.[14] Subsequently, with gathering momentum, we are now hurtling into a more advanced, frantic state of affairs.

Two basic qualities define the current episode. First there is something truly revolutionary in universal simultaneity, the access just about anywhere (that is, at the national or even global scale) to identical messages at a given instant.[15] The pioneer technology in this realm has been radio, of course, as local stations materialized, then were absorbed into nationwide networks during the 1920s. A few years earlier, effective, if not literal, simultaneity came into existence with a film industry dependent on distribution and projection of identical copies of feature films, shorts, and newsreels in theaters from coast to coast on roughly the same dates. The next momentous step came with the bursting forth of televi-

---

[14] In his *News in the Mail: The Press, Post Office, and Public Information, 1700–1860s* (Westport, Conn.: Greenwood Press, 1989), Richard B. Kielbowicz reveals how little was previously known about the subject and how rich the rewards of further research can be.

[15] The arrival of the cordless telephone as well as the telephone-equipped auto and the popularity of telephones in the lavatories of hotel rooms and private residences has meant a further contraction of truly private space. Furthermore, individuals who have reason to worry about the security of classified information have learned how difficult it is to find locations that are absolutely invulnerable to electronic eavesdropping by industrial or military spies.

sion from the laboratory by the late 1940s and the eventual acquisition of receiving sets in virtually every household and many public places, shops, offices, and even vehicles. The inevitable national hookups appeared as the radio networks shifted their priorities and cashed in on the new technology. The territorial reach and penetration of this audiovisual medium has expanded greatly in recent years with the arrival of cable TV and satellite transmission, developments that have made telecasting genuinely planetary in scope. The videocassette, whose popularity has become epidemic, is creating further options for transforming time as well as space.

At the same time, the rapid adoption of facsimile communication systems (fax) has produced virtual simultaneity for all manner of personal and business messages, while the proliferation of overnight delivery services for parcels and letters has meant a decided compression of space-time and a sort of simulation of simultaneity. But perhaps the ultimate in the collapse of both time and space has come with the approaching perfection of remote-sensing technologies. We are now able to view the entire land and water surface of this planet, or any sliver thereof, in real time or retrospectively via recorded images, using either visible light or other bands within the electromagnetic spectrum. Although the potentialities of the new medium are still far from being fully realized, we have become familiar with the meteorological applications, while exploitation for military, geological, and biological exploration and surveillance is in full swing.

A second fundamental aspect of the new geography of communications has been the interplay between two contradictory terms, a dialectical dance involving powerful forces of centralization and the opposing newer possibilities for personal expression, for distributing messages to select, not necessarily local, audiences. In this yin and yang encounter, both relatively mature and the most up-to-date technologies are being used. Thus such relatively venerable technologies as ham radio transmission, citizen band radio, and the mimeograph machine persist as media of choice or necessity for a modest segment of the communicating public. On the other hand, newer devices, such as xerography, computer hookups and mail, desktop publishing, and the narrowcasting of community and other specialized television programs offer individuals and small groups opportunities to make themselves heard and noticed.[16] The logical culmination of this trend is the display of special signs on one's person or real estate, as we shall note presently.

The foregoing development—that is, an upsurge in grassroots types of communication—may well be in reaction to the growth of oligopoly, of the control of mass information flows within fewer and larger corporate channels. Thus instead of having several daily newspapers or even a minimal pair of morning and evening competitors, most American metropolises today must make do with a single newspaper (heavily weighted

---

[16] The implications of community exhibitionism are discussed in Wilbur Zelinsky's "Where Every Town Is Above Average: Welcoming Signs Along America's Highways," *Landscape*, 30, No. 1 (1988), 1–10.

with syndicated features), and that, in turn, is most likely owned and managed by a national or multinational firm. The more successful of such enterprises have found it to their interest to engage in every profitable mode of vending information and entertainment, and often other goods and services. The result is a handful of conglomerates dominating news-papers, magazines, trade and textbooks, films, television programs, video-cassettes, and audio tapes within a wide array of countries (the ultimate proprietor of this volume being a case in point). Then too, despite what-ever legal niceties may preclude candor, the federal government, like its counterparts in other lands, has found methods for managing news and information, to direct attention to certain issues and divert it from others, to manipulate public opinion so as to buttress official policy. The general picture that emerges is that of a few Goliaths facing millions of Davids in a rather uneven contest over the management of information, and thus hearts and minds. Both camps are scoring points, but one more effectively than the other.

Unfortunately, work on the geography of communications has been woefully inadequate, so that our knowledge of the cultural implications of even the first (that is, nineteenth century) phase of its modernization is still rudimentary; and we can only speculate about the reshaping of our cultural landscape resulting from the powerful newer technologies that have just begun to assert themselves fully. The only certainty is that the impact has already been profound and may become even more so.

In the related realm of transportation,[17] recent change has been less startling. Instead of revolutionary innovations, we find the working out, the logical consummation of technologies available some decades ago. We need not concern ourselves with freight traffic since no basic changes are in evidence. The personal transportation situation is somewhat differ-ent, especially because of its critical role in transforming settlement forms and dynamics. If the airlines have gained a virtual monopoly of mass long-distance travel at the expense of passenger trains and intercity buses, the most visible result has been to spawn new office and commer-cial complexes, veritable mini-cities, but only in and near the larger met-ropolitan terminals. On the other hand, it is the personal vehicle that is radically redefining the geometry and life patterns of town and country-side. In part, this is the outcome of ever-greater numbers of automobiles, trucks, and other conveyances being relied upon for basic transporta-tion—to and from work, school, church, shops, kith and kin, and basic services—as public transit systems deteriorate and the two-car (and two-job) household becomes the norm.

A major component in the restructuring of both mobility and the settlement fabric (and the trucking business as well) has been the Inter-state Highway System and similar limited-access roads. Especially potent in this regard have been the beltways, or ring highways, encircling, and

---

[17] With the trivial exceptions of smoke and drum signals, semaphores, and carrier pigeons, communication was coincident with transportation until the introduction of tele-graph systems in the 1840s; and even then the new medium of transmitting messages was closely wedded to the railroad for many years.

sometimes disemboweling, many of the larger and even some of the smaller American metropolises. The net effect of these tens of thousands of miles of high-speed ribbons of concrete has been to widen considerably the daily, or weekly, action-space of a good deal more than half the national population, to offer much greater discretion as to where to live, work, shop, or recreate. It has also meant that not only have Americans been able to maintain their traditional phenomenal mobility, almost certainly the greatest in the world, but that they also are now migrating longer distances than ever before. In addition, the reduced friction of distance is also reflected in a substantial boom in recreational and tourist travel, and in the number and variety of second homes, motor homes, trailers, campers, snowmobiles, dune buggies, all-terrain vehicles, and even motorboats. The landscape consequences are enormous.

Transnationalization of Culture

A general cheapening, at least until lately, in the costs of moving both information and human beings and the diversification of means for doing so are essential factors in one of the most striking and fateful developments in all of human history: the transnationalization of culture and society (a topic introduced gingerly in a preliminary way in an earlier passage [pp. 86–87]). The shortest definition of this neologism would be the long-distance sharing of cultural items, along with social contacts, above a certain critical threshold. But such a bare-bones statement calls for some fleshing out. The sharing in question involves not just the elite stratum of a society—for many members thereof have indulged in cosmopolitan activities for several centuries—but rather a reasonably representative cross section of the entire populace. And the objects, ideas, and attitudes being shared, and being conveyed hither and yon at a rapid clip, may be either selected elements of the culture of distant places or the standardized components of an emergent world culture, or both. Furthermore, the process is qualitatively distinct from the relatively leisurely diffusion of traits, that venerable mechanism that has operated for many millennia and has been so lovingly studied by geographers and anthropologists, whereby gradual cultural change can and does materialize. Just how deep or long-lasting the effects of this new mode of societal evolution will be is still much too early to say.

As already suggested, recent advances in telecommunications and transportation have greatly facilitated this vast new commingling of formerly isolated pools of cultures and communities, a phenomenon to which the United States has been a leading contributor. Related to, indeed reliant upon, these new technologies are several developments that have aided and abetted transnationalization. Prominent among them is the enormous expansion in recent decades of international tourism. Again, it must be stressed, this is exotic travel for the masses, not the exclusive prerogative of the privileged few, as was true until the 1940s. The knowledge and human contacts gained by the tens of millions of tourists to and from America and other lands may be considerable. If tourism has become a truly universal activity, the United States does claim the lion's share of another variety of international traffic that does have some cultural fallout: the stationing of military personnel and their dependents at scores of installations around the world.[18] Whatever cultural contagion may occur as a consequence is not entirely trivial, but it is mostly one-way given the insularity of Americans abroad.

At a deeper level, presumably, the informal education acquired by those hundreds of thousands of college and some secondary-school students venturing abroad for study, travel, and other forms of excitement, for periods of months or years, has been transformative. Similarly, we now have a respectable corps of scholars, technicians, and other professionals circulating or migrating permanently across international boundaries. In this connection, it is worth noting that before the 1950s the American geographic discipline was staffed almost entirely by home-grown individuals. Today our ranks are richly diversified by many visitors or fully transplanted scholars from other lands; and, in return, a fair number of American geographers have found positions aboard.

A more transient but nonetheless effective mode of exchanging influential persons—carriers, so to speak, of cultural viruses—has been the international conferences held by all those many organizations that embrace the entire universe of human endeavor. From a mere handful of such events in the past century, this type of global camaraderie has flourished to the point where approximately 10,000 meetings are taking place annually in the 1990s. The United States currently ranks first among the host countries, having recently overtaken former leaders France and Great Britain. Many of these multinational events are staged by associations, mostly nongovernmental, that are self-consciously, often aggressively, transnational in character. They encompass every imaginable area of interest: science, technology, law, politics, religion, the arts and humanities, popular culture, sport, hobbies and all other forms of recreation, philanthropy, and every variety of social uplift. World's fairs, international expositions, and trade fairs of all kinds also contribute to global interchange of ideas and attitudes as well as material gear.[19]

Alongside and intertwined with all these many channels for transnational dealings, we have, of course, the multinational corporations. In

[18] The United States is hardly alone in such overseas enterprises. At various times in the recent past as well as now, British, French, Chinese, Soviet, and Cuban forces, as well as those serving under the United Nations banner, have been garrisoned in foreign nations; but during what passes for peacetime, no other country comes close to competing with America in the magnitude of its offshore military establishment.

[19] In his "The Geographical Distribution of Meetings throughout the World," *Transnational Associations*, 36 (1984), 142–59 and other writings, Gian Carlo Fighiera provides an excellent overview of both the geography and history of the phenomenon, as does the earlier Ejler Alkjaer and Jørn L. Erikson, *Location and Economic Consequences of International Congresses* (Copenhagen: Harck, 1967). We have as yet no serious attempt at dealing in depth and breadth with the larger subject of the transnationalization of society and culture; but for useful introductory statements, see James A. Field, Jr., "Transnationalism and the New Tribe," *International Organization*, 25 (1971), 335–72; Johan Galtung, *The True Worlds: A Transnational Perspective* (New York: Free Press, 1980); and Alex Inkeles, "The Emerging Social Structure of the World," in Harold D. Lasswell et al., eds., *Propaganda and Communication in World History. Volume III: A Pluralizing World in Formation* (Honolulu: University Press of Hawaii, 1980), pp. 452–515.

many respects, they are becoming *the* leading players in the conduct of the world's business today as they dominate the key industries, including manufacturing, banking, information and entertainment, retail trade, and energy. With only vestigial allegiance to the country of origin or headquarters location, the firms in question strive to standardize their operations, services, and products throughout the world with only the mildest pretence of regional or national particularity.

All these mechanisms and forces are giving us a world that begins to look qualitatively different from anything earlier generations could have known. If there has always been a certain receptivity to the exotic in the elite realms of culture—in the higher arts, in luxury goods, architecture, and the like—the ongoing change, one occurring with special vigor within popular culture, is affecting virtually everyone in all classes. And, of course, popular culture is itself a product of modern times, something that could not exist in the absence of modern means of production, distribution, and communication. The end result of the transnational imperative for our greater society has been twofold: (1) The transfer of the local and peculiar from zone of origin onto the world scene; and (2) a worldwide standardization of all manner of objects. Eventually, as what was once regional and special becomes widespread, it too becomes denatured and undifferentiated, so that the two processes just noted merge into a single phenomenon.

A few illustrations may suffice. The Chinese cuisine, once confined to its homeland and the overseas Chinese, has entered the restaurants, grocery shops, kitchens, and bookshops of the world, or a large section thereof, simplified and altered, to be sure, but nonetheless still recognizably East Asian.[20] Even though the more refined versions of such cookery are still the province of the cognoscenti in such countries as the United States and Canada, (naturalized) Chinese items have become part of the standard daily or weekly fare of ordinary folks. In similar fashion, American jazz, originally the property of a small fraction of the inhabitants of a limited section of America, has spread not only to the rest of the country but throughout the world, especially in its latter-day incarnation as rock music. It is now part of a planetary youth culture and bears little reference to regional personality. Another American invention, blue jeans—a.k.a denims or Levi's—has escaped its lowly parochial beginnings to become the standard worldwide uniform of any youth with the purchase price. Another case in point is the recent evolution of golf from an eccentric Scottish pastime, transcending both spatial and class barriers, to become the passion of millions around the globe. Other examples abound in the realms of costume, dance, hairdo, cosmetics, athletics, and furniture.

But much of the physical apparatus of our transnationalizing world is both recent in genesis and locationally nondescript. Thus our mass-produced vehicles—the "world car" being a prime exhibit—have become indistinguishable from one country to another. And an aspatial sameness

---

[20] The only effort to date to treat the geography of the interethnic sharing of ethnic foods has been Wilbur Zelinsky, "The Roving Palate: North America's Ethnic Restaurant Cuisines," *Geoforum*, 16, No. 1 (1985), 51–72.

is becoming the rule for much vernacular and commercial architecture, including hotels, apartment and office buildings, single-family homes, airports, and shopping centers and for electronic gear, clothing, furniture, comic books, and various plastic gadgets. Indeed, the process of manufacture and assembly, as in the case of computers or major motion pictures, can involve operations and personnel from and in a good many different lands.

The United States has played an especially pivotal role in this obvious homogenization of a seemingly shrunken world. Once upon a time, this country was an importer of cultural goods, including religions, literature, the plastic arts, furniture, architectural styles, and fashionable clothes, along with basic tools and technologies, domesticated plants and animals, and abstract ideas, political and otherwise, with a minimum return cargo. Today the situation has been sharply reversed. In fact, if any single nation is the principal actor in the drama of transnationalization, that honor goes to America. Although the credit, or blame, must be shared with other Anglophone lands, American influence throughout the world, in all its many ways, has been the principal factor in making English the world's universal tongue, our de facto planetary lingua franca. Aiding and abetting this linguistic coup have been American books, magazines, comics, films, television programs, whether in translation or the original tongue, (not to mention T-shirts, inscriptions, and commercial logos) that blanket the globe. In fact, it was Hollywood that effectively originated, packaged, and exported the basic formulae for the motion picture, that irresistible, ubiquitous new art/entertainment medium of the modern age, thereby greatly accelerating the universal Englishing of our verbal life.

And, similarly, American enterprise is responsible for the world pandemic of TV sitcoms, game shows, and other such delights. As already noted, American pop music, often retaining the original lyrics, has wandered far and wide, ignoring the barricades of distance, borders, and ideologies. Within the world of foodways, the American hamburger, hot dog, breakfast cereals, chewing gum, and cola drinks are being consumed in the most unlikely of places, along with the generic formula of the fast-food emporium. In addition, the United States has spawned the skyscraper, prefab construction, motels, and shopping malls, all of which have taken root in virtually every susceptible corner of this planet. This country has also devised basketball, one of those three essentially modern games (along with soccer and tennis) that claim the devotion of hordes of fans in scores of nations.

American activities abroad have had their impact upon some of the central elements of foreign cultures, not just in the sphere of fun, games, and consumer frills. As noted earlier (pp. 63–64), there is, most notably, the outreach of American churches. The missionary efforts of such indigenous groups as the Mormons or the Jehovah's Witnesses may be especially hard to ignore, but many other Protestant denominations, along with the Roman Catholics, have exported the Christian gospel, along with a bulging kit of American cultural items, to missions, schools, hospitals, and other establishments in an amazing array of locations in virtually every country beyond the national borders. It is a phenomenon that began early in the nineteenth century, took off vigorously thereafter, and still shows few signs of abating. Other "Western" Christian communities have been active in missionizing the "heathen" world, of course, but none on such a geographic or quantitative scale as the American program. Parallel to such quasi-nationalistic and overtly religious operations, Americans have also been in the forefront of philanthropic endeavor, whether in the shape of the Peace Corps or under the auspices of countless private agencies or even individuals.[21]

Even more pervasive than the inroads of American religion on the psyche and behavior of communities elsewhere has been the crucial way this nation's political innovations have shaped the evolution of nationalistic practice in so many other lands. (And if it is fair to regard nationalism as a civil religion—indeed, the dominant form of worship today—then we are still discussing religion.) The list is impressive. As already intimated, the United States was the first country to acquire and acclaim a national hero, to celebrate annual national holidays, and to create a capital city on a virgin site. It was on American soil that the very first national cemetery (Gettysburg, 1863) came into being, and a few years later the world's first national park. All these notions were adopted eagerly by a great many foreign clients. However, not every peculiarity of American political life has proved to be readily exportable, notably an almost pathological flag fetish. On the other hand, the American Constitution, which has been so widely studied and plagiarized, has had a profound effect on many new republics; and we have contributed at least three men (Washington, Jefferson, and Lincoln), arguably more than any other land, to the most exclusive of pantheons: public figures who are universally revered.

The process of transnationalization is complex and far from complete. Indeed, as we look about in the midst of so much confusing change, it is difficult to envision just what sort of world it is leading to. What is obvious is that there is much give-and-take going on among the peoples of the world via multiple channels and at a level far surpassing anything that could have been imagined in the past. It is also clear that no giver has been more energetic than the United States, nor any other land equally receptive to what the rest of the world has to offer. The moral, then, is that the cultural geography of America is, increasingly, that of the entire world; and, conversely, that all those other countries are in the grip of Americanization or, more strictly speaking, that they are partners in a worldwide collusion in which the United States still retains a vanguard position. We can no longer divorce the study of America from that of non-America. World culture and American culture have become inseparable.

Restructuring of Sensibilities and Places

Although the United States and the larger world to which it is so inextricably tied have changed quite remarkably in recent years, seemingly bedrock values and cultural axioms do remain. Like everything

---

[21] No one is ever likely to treat the topic in greater depth and detail than has Merle Curti in *American Philanthropy Abroad: A History* (New Brunswick, N.J.: Rutgers University Press, 1963).

else, including ocean basins and mountain ranges, they too must gradually mutate and evolve; but the identification of four central attributes of America's national character discussed earlier in this volume—individualism, the cherishing of mobility and change, a mechanistic vision of the world, and a messianic perfectionism—are still very much with us. Furthermore, despite all the inroads of transnationalization, our national community retains much of its distinctiveness, as do other national or ethnic groups. But in the following pages I wish to focus on modulations of traditional patterns, the ways in which dynamic forces, internal and external, are altering collective perceptions, sensibilities, place-linked behavior and attitudes, and the cultural landscape in general. I do so rather diffidently because it is necessary to touch ever so briefly—sometimes in only a single sentence—on substantial topics that deserve, and have often received, extended monographic treatment elsewhere.

Some of the changes and determinants are quantitative and readily countable. The simple fact of growth in aggregate population, from 208 million residents in 1970 to something over 250 million in 1991 has its repercussions in all aspects of our geography. We can anticipate the current increment, roughly 0.7 percent annually, to persist for some years. However, there is a strong likelihood of Zero Population Growth or even a gentle decline at some point during the twenty-first century, a prospect that will have its psychological consequences here as has happened in various European countries. The recent decline in fertility to a value below replacement level—another phenomenon shared with all other "advanced" countries—is already having its social and economic effects. Perhaps most obvious is the substantial absolute and percentile growth of an elderly population that is relatively affluent and politically potent. These oldsters are also becoming an increasingly important factor in the reshaping of the settlement landscape. Still within the realm of the numerical, the United States also resembles other urban-industrial, or postindustrial, lands in experiencing such socioeconomic trends as an unprecedented incidence of divorce and the rapid rise in number of single-person households, amount of nonmarital cohabitation, female employment, families with two wage earners, and, although difficult to enumerate, the homeless population. Unfortunately, a long-term decline in level of voter turnout may be a peculiarly American malady.

Less amenable to counting, but nonetheless quite inescapable, are certain momentous transformations in social attitudes that, like the foregoing, transcend international boundaries. Corresponding to startling changes in sexual mores is a new openness in public discourse on such matters. Along with other academicians, geographers are now able to conduct and publish research on gay and lesbian topics and on prostitution.[22] In a parallel development, we have a worldwide women's move-

---

[22] Noteworthy among the growing geographic literature on our homosexual population are such items as Barbara A. Weightman, "Gay Bars as Private Places," *Landscape* 24, No. 1 (1980), 9–16; Manuel Castells, *The City and the Grassroots* (London: Edward Arnold, 1983), 138–72; Bob McNee, "If You Are Squeamish," *East Lakes Geographer*, 19 (1984), 16–27; and Mickey Lauria and Lawrence Knopp, Jr., "Toward an Analysis of the Role of Gay Communities in Urban Resistance," *Urban Geography* 9, No. 4 (1990), 337–52. The first book-length

ment, which has generated a rush of activity in geographic and other scholarly circles in addition to all manner of reform and controversy within every level and sector of society.[23] There has also been a salutary shift in racial attitudes and relationships in the years since World War II. Even though a deeply ingrained racism does certainly persist, still blighting the lives and minds of all too many, historic barriers that frustrated and embittered millions of non-Caucasians have been partially breached. Collectively, Americans are embarrassed by overt racial prejudice and pay at least lip service to legislative, judicial, and administrative efforts to create a level playing field in housing, employment, schooling, and the arts and sciences. The presence of blacks, Hispanics, and Asians in advertisements and various entertainment venues stands in sharp contrast to the situation a generation or two ago. There is also a new civility in the way most Americans now regard and treat the physically and mentally handicapped, as well as other disadvantaged groups.

Such new sensitivities extend to matters environmental. An awareness of ecological issues, once confined to a few scholars and congenital malcontents, has permeated the general consciousness and engendered much geographic research and teaching. Indeed, it has measurably modified consumer behavior and lifestyles, even though the considerable volume of "Green" lobbying and organizational activity in America has not enjoyed the political successes attained in Western Europe. The term "organic" has entered the basic American vocabulary, and any food or other commodity bearing the label promptly gains a solid commercial advantage.[24] In its most recent extension, this latter-day openness to the nonhuman world has given rise to the animal rights movement.

ETHNIC DEVELOPMENTS.   All of the items noted above have had their repercussions, directly or indirectly, in the cultural geography of this land, but perhaps none more telling than recent developments in ethnic matters. In an earlier passage (pp. 22–28) we discussed the crucial role of successive and varied waves of immigrants in shaping both national and regional patterns of culture. Writing twenty years ago, one could only hint at the strikingly different composition and impact of the most recent episode in American immigration history. It began in 1945 after that thirty-

---

essay on the geography of prostitution, Richard Symanski, *The Immoral Landscape: Female Prostitution in Western Societies* (Toronto: Butterworths, 1981), has met with a mixed critical reception.

[23] Three of the more useful introductions and bibliographic guides to the rapidly expanding literature on the geography of women are Wilbur Zelinsky, Janice Monk, and Susan Hanson, "Women and Geography: A Review and Prospectus," *Progress in Human Geography* 6, No. 3 (1982), 317–66; Mary Ellen Mazey and David R. Lee, *Her Space, Her Place: A Geography of Women* (Washington, D.C.: Association of American Geographers, 1983); and Suzanne Mackenzie, Restructuring the Relations of Work and Life: Women as Environmental Actors, Feminism as Geographic Analysis," in Audrey Kobayashi and Suzanne Mackenzie, eds., *Remaking Human Geography* (Boston: Unwin Hyman, 1989), pp. 40–61.

[24] Warren J. Belasco has dealt with this topic in eminently satisfying fashion in *Appetite for Change: How the Counterculture Took on the Food Industry, 1966–1988* (New York: Pantheon, 1989).

year hiatus caused by two world wars, a major economic depression, and a set of highly restrictive, now mostly defunct, legislative enactments.[25]

The magnitude of current legal immigration is impressive—roughly 500,000 to 700,000 a year. In addition, of course, there is a substantial volume of illegal entries (and exits), the extent of which is highly uncertain and subject to much academic as well as official controversy. The provenance of these new residents is dramatically at variance with that recorded in times past. Increasingly, we are accepting newcomers from Eastern and Southern Asia and from Latin America, with lesser streams from the Middle East and Canada, and declining numbers from traditional European sources. Although also heavily weighted by arrivals from the Caribbean and Latin American lands other than Mexico, the illegals include a wide variety of other folks, among them Asians, Canadians, the Irish, and other Europeans.

But it is not just the source areas that have changed and, to some degree, the places chosen for settling down. More germane to my argument is the sort of welcome accorded the newcomers. In place of the hostility and xenophobia confronting earlier non-British immigrants, we find a surprising change in general attitudes. Admittedly many incoming Mexicans, Puerto Ricans, Haitians, Iraqis, and Salvadorans do not enjoy the jolliest of times in their new domiciles. But consider the nearly 180-degree turnabout in our perceptions of recent Japanese, Korean, and Chinese immigrants or Hungarian, Czech, Vietnamese, Cuban, Iranian, and Soviet Jewish refugees, or their remarkably fast climb up the socioeconomic ladder, along with that of many individuals from Thailand, the Philippines, Pakistan, and India.

The fact that their level of education and their skills are higher than those of earlier generations of compatriots and that they are no longer dismissed as the "wretched refuse of the teeming shore" is only part of the explanation. No longer are we revolted by their weirdness; indeed, we relish their fascinating customs, foods, and crafts, and actively seek them out. In fact, if there is a problem, it is that they are assimilating too rapidly, becoming super-Americans almost overnight. In a way, this new sensibility is transnationalism brought home, tourism with minimal exertion, and certainly a basic revision of earlier American attitudes toward the strange and different. The shrewder entrepreneurs among the newcomers have lost no time in capitalizing on this mass hunger for titillating ethnicity; but it is important to note that such enterprises would have been unthinkable not too long ago.

Not only are many of the newer immigrants to the United States enjoying a measure of cultural hospitality and respect unknown previously but we also find the older hyphenated Americans bestirring them-

---

[25] Although it was comparatively only a trickle, the novel character of the new influx was already apparent in the 1930–1945 period with the flight to America of an extraordinary company of politial refugees from Germany, Austria, and other Western European countries. Jews and Gentiles alike, these scientists, scholars, writers, artists, and other creative folk greatly invigorated the intellectual and scientific life of America. One of the more informative accounts of this hegira is to be found in Donald H. Fleming, ed., *The Intellectual Migration: Europe and America, 1930–1960* (Cambridge, Mass.: Harvard University Press, 1969).

selves, remembering, restoring, and celebrating—or inventing—the ways of their forebears as they join the Ethnic Revival, a movement very much alive in Canada and various European countries as well as here.[26] The European "ethnics," that is, Poles, Czechs, Slovaks, Croats, Magyars, Jews, Ukrainians, Basques, Italians, Germans, Irish, Welsh, Greeks, Portuguese, Lithuanians, and Romanians among others, are not the only players in the game. In fact, the first great burst of ethnic revivalism was the black pride movement, beginning in the 1960s, to be followed by the arousal of the Cajuns in Louisiana and adjacent states, the Chicanos in the Southwest, the Native American movement among the scattered remnants of our aboriginal population, and some revitalization of the Polynesians in Hawaii. Closely parallel to, and simultaneous with, all this neoethnicity was the rediscovery of the fact that there are genuine, interesting regions within this country, the ramifications of which we shall explore presently.

SHIFTING SETTLEMENT PATTERNS.   Let us turn next to recent place-related developments in American life and culture. Perhaps the most visible of these are the major changes—basic transformation may not be too strong a description—going on in settlement patterns. As striking a signal as any of the emerging new order is the blending of the rural and the urban with the wearing away of meaningful divisions between the two. But if the traditional urban-rural dichotomy, along with the familiar continuum from major metropolis down to the tiniest hamlet, has lost so much of its former validity, location and population size and density are attributes we must always reckon with. Certain remote or sequestered places may forever retain the look of being genuinely rural; and only the largest of metropolitan areas can operate an international airport, a cluster of skyscrapers, a medical school combined with a teaching hospital, or a major league baseball or football team.

There is ample statistical and other evidence to support the case for urban-rural convergence in the United States. To take some of the more useful surrogate indices of general societal values, the excess of rural fertility over the urban that had been so considerable since colonial days now seems to be approaching the vanishing point. The very same thing has happened to the venerable discrepancy between the median ages of the urban and rural nonfarm population. If differences in personal income levels of the two residential categories still linger, they have grown much smaller during the past few decades and may ultimately

---

[26] It is rather surprising to discover the lack of any sharply focused study of the American Ethnic Revival in its entirety within the academic literature. What we do have are treatments of individual groups, e.g., the Chicanos, Afro-Americans, or Native Americans, or tangential discussions in general treatments of ethnicity in the United States, as in Stephen Steinberg, *The Ethnic Myth: Race, Ethnicity, and Class in America* (New York: Atheneum, 1981) and Herbert J. Gans, "Symbolic Ethnicity: The Future of Ethnic Groups and Cultures in America," *Ethnic and Racial Studies*, 2 (1979), 1–20. The subject has received much greater attention in Western Europe and Canada where we can consult Anthony D. Smith, *The Ethnic Revival* (Cambridge: Cambridge University Press, 1981) or a number of articles by Colin H. Williams, such as "Ideology and the Interpretation of Minority Cultures," *Political Geography Quarterly*, 3, No. 2 (1984), 105–25.

disappear. In similar fashion, we observe a radical shrinking of late in the differences between median values of urban and rural dwellings and in the material equipment of these structures, including such items as modern plumbing. Notable also is the thinning out of metropolitan population densities as cities sprawl farther and farther into the countryside at the same time that most rural counties have stabilized their population numbers or are experiencing a resurgence thereof. The dispersion of manufacturing—an activity once virtually synonymous with urban location—into the countryside has been amply noted.[27] Surprisingly, as defined by the Census, the percentage of the (nominally) rural labor force now engaged in manufacturing exceeds the urban value. Whatever differentials may persist between the two residential categories in terms of mortality, cause of death, marriage and divorce rates, educational attainment, or household and family characteristics are a good deal less than those to be found *within* most individual metropolitan areas. Perhaps the most meaningful of the residual distinctions still remaining have to do with social values and attitudes. Changes in those are difficult to track, but are probably diminishing.

The historical and technological factors promoting the coalescence of town and countryside are obvious enough. The first critical event was the building of a pervasive railroad system,[28] followed some years later by Rural Free Delivery, the triumphs of the mail-order industry, and the paving of all-weather roads connecting farms, villages, and cities. Mass conscription during World Wars I and II brought together and mingled tens of millions of young men (and not a few young women) from all places and strata of society, and began a certain erosion of sociocultural differentials.

But it was really not until the New Deal era and the post-1945 years that we witness truly profound change in the transactions between city and hinterland. The introduction of electricity throughout the back country that began on a heroic scale in the 1930s has had a revolutionary impact on both society and the economy.[29] Somewhat later, in the mid-1950s, the new Interstate Highway System greatly enhanced the accessibility of most populated places. But, as already noted, the arrival of advanced telecommunications may well have furnished the most powerful of solvents in hastening the dissolution of urban-rural contrasts.

In addition to the work of such lubricants as modern means of transport and communication in facilitating the flow of ideas and attitudes, the settlement tissue that is the American world has had grafted onto it a huge

---

[27] Richard E. Lonsdale and H. L. Seyler, eds., *Nonmetropolitan Industrialization* (New York: John Wiley & Sons,Inc., 1979); Howard G. Roepke, "Industry in Nonmetropolitan Areas," in Rutherford H. Platt and George Macinko, eds., *Beyond the Urban Fringe: Land Use Issues of Nonmetropolitan America* (Minneapolis: University of Minnesota Press, 1983), pp. 149–58.

[28] John R. Stilgoe, *Metropolitan Corridor: Railroads and the American Scene* (New Haven, Conn.: Yale University Press, 1983).

[29] Robert Caro recounts the personal and economic results of rural electrification in Texas quite movingly in *The Years of Lyndon Johnson. Vol. 1: The Path to Power* (New York: Knopf, 1982), 502–28.

far-flung gauze of exurban folk—again to a degree unmatched elsewhere on this planet.[30] They are those millions of persons, whether economically active or retired, who reside well beyond even the outermost metropolitan suburbs but do not derive their livelihood from their immediate environs. Instead, in both thought and deed they live an essentially urban life but within bucolic surroundings, forming a "galactic city" for whose geometry there is no forewarning in our standard texts on urban places. The metro/nonmetro "Turnaround," that reversal of historic trends, from the late 1960s to about 1980, when the number of in-migrants to nonmetro areas exceeded the flow in the opposite direction (a phenomenon occurring simultaneously in a number of other advanced countries) certainly facilitated this new urbanoid formation.[31] However, exurbanization, and thus the remaking of the countryside into something functionally more urban than agrarian or pastoral, has continued apace into the 1990s, even though the metro areas seem once again to be gaining more internal migrants than they are surrendering.[32]

In a related development, we must also recognize a revolutionary reorganization of our conventional metropolitan area with the creation of so-called Edge Cities. These are not the familiar residential suburbs but rather virtually autonomous, self-contained concretions of facilities—commercial, office, residential, recreational, social, and other—at strategic locations along the peripheries of the larger cities. They have sprouted and mushroomed at not one but several points in the outskirts of Washington, D.C., New York, San Francisco–Oakland, Los Angeles, Dallas–Fort Worth, and Boston, among other places. Their vitality has been at the expense of the historic center city, so that, in effect, the American metropolis is being turned inside out.[33]

But the phenomenon in question runs much deeper than any simplistic diffusion of things outward from the city or the easing of spatial friction via latter-day technological marvels. It is necessary to invoke one of the more central elements in the American credo. This is a nation still haunted by the Jeffersonian vision of the good life, of that magical middle landscape poised between civilization and wilderness[34]; we remain a

---

[30] Peirce F. Lewis,"The Galactic Metropolis," in Rutherford H. Platt and George Macinko, eds., *Beyond the Urban Fringe: Land Use Issues of Metropolitan America* (Minneapolis: University of Minnesota Press, 1983), pp. 23–49.

[31] Stephen E. White et al.,"Population Geography: Counterurbanization," in Gary L. Gaile and Cort J. Willmott, eds., *Geography in America* (Columbus, Ohio: Charles E. Merrill, 1989), pp. 266–69.

[32] The ongoing regional redistribution of the American population, most notably to certain sections of the South and West, is of only peripheral relevance to this discussion, but it is another indication of a changing set of sociocultural values and aspirations.

[33] For a splendid account of the history, anatomy, physiology, and implications of this remarkable new settlement type, see Joel Garreau, *Edge City: Life on the New Frontier* (New York: Doubleday & Co., Inc., 1991). Another rewarding discussion is available in John Herbers' *The New Heartland: America's Flight beyond the Suburbs and How It Is Changing Our Future* (New York: Times Books, 1986).

[34] The classic statement on the subject remains Leo Marx, *The Machine in the Garden: Technology and the Pastoral Ideal in America* (New York: Oxford University Press, 1964). Joseph S. Wood provides a definitive treatment of the particular role of the New England village in the fabrication of the Middle Landscape Myth in " 'Build, Therefore, Your Own

people nursing an innate dislike and distrust of large cities, arguably much more so than almost any other national community. It is only recently, with greater affluence and more generous options in lifestyle and location, that many of us are finally able to act out this collective daydream.

We can find eloquent testimony for this contention in a striking recent development: the ruralization of the American city. Such a rural tropism is blatantly visible in the suburban home with its lawn, yard, garden and barbecue setup, a sort of simulated yeoman's farmstead. The popularity of agrarian mementoes (wagonwheels, old-fashioned pumps, milk cans, wooden wash tubs, fake water wells, and the like) for yard ornaments and mailbox props underscores the statement. Relevant too is the widespread adoption by entrepreneurs of names for their residential developments and streets that reek of the pastoral.

But ruralization also reaches deep into the core of the city proper. This is more than the spectacle of millions of in-migrants from farms and small towns entering the cities over a span of a hundred years or more and their subsequent impact upon the ways of urban folk, a topic American scholars have not yet adequately explored. As symptomatic a clue as any to the rural yearnings among our city-born and city-bred has been the extraordinary growth in the popularity of country music in cities large and small and among all strata of the population from top to bottom. The proliferation of radio stations dedicated full- or part-time to the genre is further witness to the trend. And so too the success of various television programs exploiting the down-home rural mystique, as well as countrified bars and restaurants, especially those trendy ones featuring barbecue, that quintessentially outdoorsy gift of the rural South to the national cuisine. In the same vein, take note of the recent adoption by so many city dwellers of a particularly fetching form of rural attire: Western costume, fancy boots, oversize hat, exotic shirt, belt buckle, plus all the minor accoutrements. It is a small or culturally deprived city that does not support at least one Western clothing emporium, one generally catering also to the equine trade. Just a generation ago all those countrified pickup trucks inhabiting so many metropolitan driveways today would have been snubbed as being infra dig and hopelessly hayseed. The use of rural motifs in a number of the more effective TV commercials (concocted, of course, in such places as New York City or Los Angeles) also comes to mind. The point of citing such facts is to document a startling fact—and a likely example of American exceptionalism—that what has been going on is a complicated two-way traffic in influences. While the countryside may be becoming citified, the city is simultaneously undergoing rustication.

Such a ruralization of the American city, as mirrored by the urbanization of the countryside, may be only a single facet of an even more deepseated revision of the ways Americans relate to their habitat. J.B. Jackson, that shrewdest watcher of our landscape and its dynamics, has suggested,

World': The New England Village as Settlement Ideal," *Annals of the Association of American Geographers*, 81, No. 1 (1991), 32–50.

albeit tentatively, that we have entered a period of transition from what he identifies as Landscape Two to Landscape Three.[35] The former, which has been dominant throughout most of the nineteenth century and well into the present one, is a well-ordered state of affairs in which centralized economic and political power determines the lay of the humanized scene. It superseded the essentially medieval Landscape One, a set of "Thousands of small . . . impoverished vernacular landscapes, organizing and using spaces in their traditional way and living in communities governed by custom, held together by personal relationships." Landscape One was only locally and fitfully transplanted into North America because it was so quickly overtaken by the forces of a modernizing North Atlantic World.

Unlike Landscape One, which mixed all kinds of uses and spaces together, Landscape Two insists on spaces which are homogeneous and devoted to a single purpose. It makes a distinction between city and country, between forest and field, between public and private, rich and poor, work and play . . . it clearly believes that whatever is temporary or short-lived or movable is not to be encouraged.[36]

The character of the emergent Landscape Three, which we might identify as postmodern, is still not entirely clear except for its negation of Landscape Two and an ever so distant family resemblance to Landscape One. Fluidity, transience, improvisation, cancellation of distinctions among the ways spaces are used, a redefinition of privacy—these seem to be some of the defining attributes. Jackson was looking at the lower end of the social scale in writing about

the kind of new community that we are seeing all over America: at remote construction sites, in recreation areas, in trailer courts, in the shanty towns of wetbacks; the emergence of what we may call vernacular communities—without political status, without plan, ruled by informal local custom, often ingenious adaptations to an unlikely site and makeshift materials, destined to last no more than a year or two, and working as well as most communities do.[37]

Equally symptomatic of the new landscape dispensation are such upscale developments as the shopping mall, the pricey, security-obsessed residential enclave, or the entire tribe of Edge Cities. What is public and what is private in our self-contained shopping centers? These are places where the management goes to some pains to ban leafleting, demonstrations, panhandling, and other standard practices of the outmoded old Central Business District, but where it is hard to demarcate any sharp lines between social territory and the various shops and where open spaces are frequently sites for various civic ventures, festive events, and the promotion of worthy causes. The Edge City exists in a curious political/legal limbo (often without it own Zip code) being controlled de facto by a shadow government—that is, some private corporate entity. The arrival of a new mind-set may be read in the avoidance of rectangu-

[35] John Brinckerhoff Jackson, *Discovering the Vernacular Landscape* (New Haven, Conn.: Yale University Press, 1984).
[36] *Ibid.*, p. 152.
[37] *Ibid.*, p. 156.

larity in the newer suburbs and parks and in house designs in which inside and outside flow together and where doors and distinctions between chambers of fixed purpose have effectively vanished. Landscape Three is a new unfamiliar kind of world, one still imperfectly observed or understood, in which there is a blurring and scrambling of places, people, and functions unlike anything known in the past.

THE NEW LANDSCAPE AWARENESS.   One of the more fascinating developments in academia and the general public arena has been the burgeoning of interest in, and sensitivity toward, landscape matters. One can hardly refrain from asking, Why now? Converts to the landscape faith have been especially conspicuous within the geographic fold. This is something that could be taken for granted given the essential nature of the geographer's mission. But the question remains: Why is it only lately that we have begun to consider seriously the actual appearance and physical personality of our surroundings, or, even more importantly, to seek out the meaning of what is perceived? The recent vogue for behavioral geography and environmental perception offers a partial explanation, for it has sensitized many a younger scholar to the urgency of evaluating our environs. Similarly, the extension of professional curiosity into humanistic and social topics has opened some eyes to the visible, tangible landscape. Thus some venturesome souls have begun considering the interplay between literature and locality to the point of exploring the semi-fictional landscapes of Dickens, Hardy, Faulkner, and Malcolm Lowry, among other writers.[38] In parallel fashion, British geographer Denis Cosgrove has peered deeply into the social and geographical connotations of landscape painting from the Renaissance onward.[39] The growing respectability of social geography has emboldened such investigators as James Duncan and Peter Hugill to look into the relationships between social class and residential landscapes in particularly revealing places.[40] Still other geographers have begun, rather spontaneously, to realize the value of decoding the landscapes generated by popular culture.[41]

[38] Although only incidentally geographic, the best general treatment of the interaction between place and imaginative literature may be Leonard Lutwack, *The Role of Place in Literature* (Syracuse, N.Y.: Syracuse University Press, 1984). For the views of geographers, see J. Silk, "Beyond Geography and Literature," *Society and Space*, 2 (1984), 151–78 and L. A. Sandberg and J. S. Marsh, "Focus: Literary Landscapes—Geography and Literature," *The Canadian Geographer*, 32 (1988), 266–76.

[39] Denis E. Cosgrove, *Social Formation and Symbolic Landscape* (Beckenham: Croom Helm, 1984).

[40] James S. Duncan, Jr., "Landscape Taste as Symbol of Group Identity: A Westchester County Example," *Geographical Review*, 63 (1973), 334–55; Peter J. Hugill, "English Landscape Tastes in the United States," *Geographical Review*, 76 (1986), 408–23.

[41] The literature falling within this genre is most diverse, but some representative samples would be Daniel D. Arreola, "Mexican Restraurants in Tucson," *Journal of Cultural Geography*, 3, No. 2 (1983), 108–14; James R. Curtis and David M. Helgren, "Yard Ornaments in the American Landscape: A Survey Along the Florida Keys," *Journal of Regional Cultures*, 4, No. 1 (1984), 78–92; Melvin E. Hecht, "The Decline of the Grass Lawn Tradition in Tucson," *Landscape*, 19 (1975), 3–10; and Richard D. Hecock, "Changes in the Amenity Landscape: The Case of Some Northern Minnesota Townships," *North American Culture*, 3, No. 1 (1987), 53–66.

Even more to the point, both academics and amateurs in other disciplines have enlisted in the landscape movement. New journals, newsletters, and conferences devoted to landscape topics began to multiply only the day before yesterday, so to speak. Plainly more than mere coincidence is at work. The almost simultaneous infection of so many denizens of Academia with the landscape bug is a sign that something larger may be brewing. In addition, related developments outside our ivied halls could indicate some subterranean shift in the order of things. Note the proliferation of popular shelter, gardening, and travel magazines, accompanied, of course, by an enormous surge in both international and domestic tourism. The fact that virtually everyone and his uncle has become a photographer is also symptomatic of the newer sensibility. Consider also the flowering of the motion picture as the most expressive of the popular arts, this being a medium that heavily exploits locales and, however superficially or misleadingly, makes us aware of alternative landscapes. Going further afield, reflect on the near-universality of planning and zoning in town and country and, especially, the growing incidence of sign, tree, and appearance ordinances and of beautification projects. Furthermore, I see a linkage, in both dating and ultimate causation, between the trendiness of landscape activity and an environmental movement that refuses to go away. And would it be too far-fetched to speculate about the temporal coincidence between the landscape vogue and our robust historic preservation industry—or the entire nostalgia binge for that matter?

I must also urge the reader to think about another novel, unprecedented phenomenon that has been going on, literally right under our noses, within recent years and is thus suspiciously concurrent with all of the above, a veritable eruption of exhibitionism on the part of individuals and communities. This public flaunting of all manner of verbal and nonverbal messages takes many forms. Although firm documentation is elusive, all available clues suggest that the planting of hillside letters and welcoming signs on the approaches to our towns is a recent, briskly ongoing thing.[42] An American would have to be totally blind to her surroundings of late not to notice that epidemic of T-shirts bearing an incredible assortment of statements, many of them unique to the wearer. Closely akin to that practice is a collusion between businessman and consumer known as "designer clothes," whereby millions of us saunter about with jeans, shirts, shoes, caps, or whatever emblazoned with trade names or prestigious emblems. This would have been outlandish a generation ago. Similarly, there has been that remarkable rise in the wearing of buttons on jackets, shirts, sweaters, and headgear, said objects conveying every conceivable (and many an inconceivable) form of utterance and artistic expression. Not too many elections ago, buttons were something you pinned on only during campaigns.

Who can deny that the compulsion to attach witty or pugnacious bumper stickers or personalized license plates on our vehicles has been

---

[42] James J. Parsons,"Hillside Letters in the Western Landscape," *Landscape*, 30, No. 1 (1988), 15–23; Wilbur Zelinsky, "Where Every Town Is above Average: Welcoming Signs Along America's Highways," *Landscape*, 30, No. 1 (1988), 1–10.

spreading like wildfire the past two decades? Although, once again, quantitative proof may be lacking, my impression is that the (largely surreptitious) practice of painting graffiti on walls, bridges, and public vehicles (an unintended consequence of the invention of the spray can) has generally been on the increase in recent years, despite police efforts at suppression. In any case, there can be little doubt that outdoor mural art, whether or not officially sponsored, is on the upswing. Conspicuous among such displays are items celebrating ethnic themes. Although compositions by Mexican, Puerto Rican, and other Hispanic artists, both amateur and professional, may be the most numerous, the black and East Asian communities are also well represented within the genre. The artistic quality of the better examples is truly outstanding. And I must refer again to the efflorescence of personal whimsy in the signs and decorations in the yards, porches, windows, and other vantage points in homes and vacation cottages. The point of this discussion is that so many more of us have begun to take some heed of what is immediate and visible, and that, in impressive numbers, we are wittingly, if often ephemerally and on a micro scale, doing something to modify the scene.

It would be foolish to imagine that the landscape-related behavioral syndrome (to coin an awkward expression) just sketched is simply the product of happenstance. A more plausible view is that the grander forces of social history, which are so closely intertwined with technological change, lie behind our altered sensibilities and dealings with the material worlds as perceived or imagined. Once again, we have a dialectical situation of interacting, reversible cause and response. The matters under discussion represent a set of individual and collective reactions to the convulsive and bewildering economic, social, and in the broadest sense, political transformation of the human realm, a process with the most uncertain of culminations.

But these may be lofty, unconvincing generalities. Suppose we concentrate on what the new technologies of communication and transportation have meant for our cohabitation with the landscape—and examine the darker reverse side of the coin exhibited above. The present-day actuality is that nearly all of us inhabit a world of illusion and manipulation with only the most restricted opportunities for perceiving and revising its character. Some specifics are called for. Isn't it true that more and more of us spend an increasing share of our days and hours in carefully programmed situations in a world of no surprises, and one snugly insulated from the natural elements? Our environmental encounters have been pretested for us. In fact, such indulgences have been preselected, precut, precooked, prepackaged, prechewed, and predigested. Consider how many Americans reside in sequestered apartment complexes surrounded by walls, guards, and tasteful greenery and flowers, places equipped, at a minimum, with pools, tennis courts, saunas, and hobby rooms.

Even more comprehensive are the delights of retirement villages; their guests need never stir outside those ever so thoughtfully managed facilities. Our briefer sojourns in shopping malls and theme parks, those risk-free micro-worlds of make-believe, offer complete protection from the elements and those naughty characters lurking in that jungle out there. For many men and women, the most intimate rendezvous with Mother Nature is a matter of chugging along the fairway of a scientifically engineered golf course. But there is always the option of playing a video-tape of ocean surf or waterfalls. A large fraction of the work force earn their keep in often windowless cubicles within high-rise structures or shops where temperature and lighting are monitored by machine, where air is constantly recirculated, and Muzak soothes the soul. They may commute to work in mass-produced steel cocoons paying attention to little but traffic and road conditions while glued to their Walkman or radio.

When we venture forth to seek travel thrills, we can do so painlessly, quite possibly along Interstates with their standardized engineering and landscaping, pausing from time to time at totally predictable, look-alike franchised filling stations, restaurants, and motels, and, if we fly, we shall probably watch the in-flight movie rather than try to make sense of the cloudscapes and landscapes visible through the window. When we reach the resort hotel or board the cruise ship, we can, in essence, switch to automatic pilot, just going with the flow of predetermined excitements. When we visit unfamiliar places, unless we are herded with a group tour, it is awkward to do without an omniscient guidebook that tells us in no uncertain terms what is worth looking at and how—and, by implication, what is unworthy of our gaze—where to eat, sleep, and disport ourselves. (How difficult now to imagine the premodern world, one bereft of guide-books!) Then, to avoid any faux pas, we have at our disposal all those historical and scenic markers, the easily negotiated nature trails winding through a tamed wilderness, with labelled trees and plants and benches for the weary, not to mention those scenic outlooks with ample parking and coin-operated telescopes pointing in the approved directions.

Perhaps the ultimate in protective layering is in our art museums where you can rub up against the great outdoors by admiring framed and labelled depictions of wild places hung on windowless walls while you scan the catalogue, listen to tape cassettes instructing one how to experience aesthetic rapture, or are shepherded from canvas to canvas by some clever docent. For those who blanch at the rigors of sightseeing or museum-hopping, there is a comfortable alternative: those packets of color slides for sale at museum and gift shops everywhere. Eventually, when we have passed on to our reward, our loved ones can pay their respects at a final resting place in a neatly manicured, civilized cemetery, where, again, there will presumably be no environmental surprises for all eternity.

Admittedly I have been caricaturing the actual state of affairs, but only slightly, for the sake of argument. Still it is difficult to deny a strong drive toward (a) the homogenization of human experiences in our advanced societies, and (b) a concomitant trend to sever us from meaningful contact with environmental and social realities. But, as social history unfolds, every force seems to breed a counterforce, every development its counterdevelopment. The upshot of the situation, with respect to landscape matters, is that, individually or locally, we have been reacting to those macro forces, carrying on a kind of guerrilla—or is it rearguard?—campaign, in the diverse ways described previously, against immense socioeconomic trends that not even the most determined of governments or other organizations can do much to stem, divert, or modify. Is it any wonder, then, that Landscape Three has been entering the scene?

According to the (J.B.) Jacksonian definition, landscapes are artificial compositions, and one of the more relevant lexical definitions of the term "composition" may be "the disposition of the parts of a work of art." When such a disposition comes about, as it often did in premodern times, through some unspoken, unwritten compact as to how life should be lived and what things should look like, the outcome could be harmonious, even attractive. But today, in a highly dynamic world—in which no national community is more dynamic or individualistic than the American—we seem to have jettisoned the old understandings. A head-on clash between two opposing sets of forces—those of centralizing conformity and standardization as against the lively, but much weaker, impulses of personal and local self-realization—has created a set of jarring incongruities. Thus in Landscape Three we have places where old and new, good and bad, things stable and evanescent, cherished and despised, like and unlike, sit cheek by jowl, and visual chaos reigns.

Such dissonance is most striking along, but hardly limited to, our newer highway strips. Taken one by one, the buildings may be acceptable, some even stunning; even the much maligned roadside billboards can be defended if we inspect them individually. Indeed, some would be assets to museum collections. It is as assemblages, as compositions, that these things disappoint. The blame is not monopolized by any single party, not by architects, developers, engineers, zoning officials, property owners, or the citizenry at large, although none of these constituencies is wholly innocent. We must look to the totality of our social-economic-political-cultural system to find the basic causes—and potential therapies.

Regional/Ethnic/Historic Revivals

A heightened landscape sensibility or activism is only one of the potential cultural or psychological responses to the disappointments of our contemporary world and the threat of an even bleaker future. In flailing about, singly or collectively, for firmer, more authentic modes of anchorage, many of us have tacitly rejected the faith that ruled supreme in the Western world for the past two or three centuries: the doctrine of perpetual progress. In effect, the party of the past has outvoted the party of the future. We have turned to look backward in time toward genuine or contrived ancestral communities, so that historic preservation in all its many manifestations has become a growth industry and ethnicity is enjoying rejuvenation. In spatial terms, the safest haven appears to be located somewhere midway between the cold, faceless bureaucratic state and the incertitudes of kith, kin, and hearth—in the renewal of regional identity. And, of course, all three varieties of nostalgic questing are intimately interrelated.

Let us consider some of the specific ways these revivals have entered the contemporary scene. As striking as anything has been the boom in new journalistic efforts catering to regional and ethnic appetites. We now have hundreds of latter-day regional and metropolitan periodicals. The circulation of the more profitable enterprises, such as *Southern Living, Sunset, Yankee, Arizona Highways,* and *Midwest Living,* runs into the hundreds of thousands, no mean feat in an era of cutthroat competition in the magazine racks. A number of university presses (and even some commercial firms) have found it advantageous to become explicitly regional in their publishing strategy, and regional institutes and think-tanks have been sprouting of late in various localities around the country. Is it merely coincidence that during the 1980s the geographic avante-garde was dazzled by the revelation that regional description and analysis could yield an abundant intellectual payoff? This after three decades of disdaining the very idea of regional study. But the current vogue is original in spirit and practice—a case of new wine in old bottles.[43] Canny business people are learning that it pays to work the regional angle in their advertising and marketing schemes—a form of applied cultural geography. Even such an international behemoth as McDonald's has adopted regional motifs in its interior decor; and the plugging of regional themes has become far from uncommon in shopping malls, theme parks, restaurants, trailer camps, and gentrifying urban neighborhoods.

The popularity of traditional regional building styles in recent upscale suburban architecture is too widespread to ignore. Admittedly, this is not something that began just yesterday. Some time ago the once highly localized Cape Cod house migrated from New England to become one of the more conspicuous elements on the American landscape coast to coast, as did the bungalow, originally indigenous to California, and roughly during the same period. There has been no break over time in the ways homes and barns are designed in southeast Pennsylvania, just a gentle evolution. But when we look at various spots in Acadiana (southern Louisiana), Quebec, California, the suburbs of almost any Deep Southern city, just about any of the burgeoning towns of the Southwest, and many another locality, we find recent, deliberate efforts to mimic and update the old indigenous building styles, to use the home or commercial structure as a badge of regional belongingness. The recycling of venerable building types is not entirely regional in character, since we seem to be experiencing a nationwide vogue for neo-Victorian residences; but, at the other extreme, this archaeological approach to domestic building can be quite place-specific, as in the suburban homes in Austin, Texas, modeled after the distinctive German-American structures of nearby Fredericksburg.

---

[43] The sheer volume of the recent highly cerebral writings by geographers on region and "locality" is impressive. Among the better introductory, or programmatic, statements, we have Doreen Massey, "New Directions in Space," in Derek Gregory and John Urry, eds., *Social Relations and Spatial Structures* (London: Macmillan, 1985), pp. 9–19; Mary Beth Pudup, "Arguments within Regional Geography," *Progress in Human Geography,* 12, No. 3 (1988), 369–90; and Nigel Thrift,"For a New Regional Geography 1," *Progress in Regional Geography,* 14, No. 2 (1990), 272–79.

There are other factors beside the rediscovery of ethnicity that can help account for the current surge in regional sentiments, in effect the restoration of culture areas as they were or, rather, as they should have been. There are connections, I suspect, with the advent of professional planning, especially at the regional scale. Regional planning is a movement that came into its own during the New Deal period and has persisted since in many venues, but there were interesting stirrings among some visionary thinkers as far back as the 1920s. Concurrent with the acceptance of regional planning during the Great Depression was a lively interest in the folk cultures of the land, which undoubtedly accompanied the popularity, on an unprecedented scale, of regional fiction, drama, and art. The celebration of regional diversity in both folk and sophisticated media—a kind of overture to the current episode—may have lost momentum in the postwar period, but it has most certainly rebounded.

Although the old foreign-language press has been languishing for some time, the new immigration has fostered a fresh generation of such periodicals. But more to the point has been the launching of Afro-American, Jewish-American, Italian-American, and other English-language magazines aimed at third or nth generation foreign stock. Language maintenance or revival has not been as burning an issue in the United States as in Canada, Wales, Ireland, Belgium, Iraq, Sri Lanka, or portions of Spain, France, and the Soviet Union, although some Americans are flustered by the vitality of the Spanish language, which has turned major portions of at least two more states (Texas and Florida, in addition to the previously hispanicized New Mexico) and many urban neighborhoods into bilingual tracts.[44] Some hyphenated American groups, most notably Polish, Japanese, and Jewish, have unabashedly mounted formal efforts to teach the ancestral tongue in public schools and colleges or in after-hours, or Saturday, classes. Furthermore, an increasing number of radio stations devote all or part of their broadcasting hours to programs directed toward the so-called minorities, using the foreign language or English, or some combination thereof. Hispanic and black stations are only the most numerous within this category, while at the other extreme are single weekly programs in Cajun French and Navaho. But all such broadcasting, along with similar television programming in some of the larger markets, not only helps maintain the language (or dialect) but also reinforces cultural identity and pride.

The vigor of the ethnic revival and kindred regional sentiments registers dramatically in the recent booming ethnic and regional festivals, an activity, indeed a veritable industry, that is reaching the megabuck

---

[44] There is exquisite irony in the way that English First and similar drives have entered the political agenda locally and even nationally when one considers that of all the languages in the world English is the least endangered. Apart from the fact that educators, among other concerned citizens, have fretted perennially about the reluctance of Americans to learn foreign languages, there is an obvious eagerness of the new immigrants to acquire and use English. The current distress among some people is reminiscent of the forebodings expressed a century ago when the country was seemingly being overrun by immigrants. The interesting difference today is the intention among many newcomers to remain bilingual.

level.[45] Such celebrations are not entirely new, of course. Mardi Gras has been observed in traditional style in small-town Louisiana, as well as New Orleans, for many years, and so too the St. Patrick's Day parade in major cities. Fourth of July and Memorial Day ceremonies, now quietly fading from the scene, frequently had local or regional overtones, while country and state fairs, which have been with us for some time, still retain their customary robustness.

But, quite apart from these rather timeworn practices, we are now experiencing a whole new breed of festival, one with many subcategories, but all involving the hawking of merchandise amidst a wide range of other activities. Thus we have events saluting the arts, music, sports, crafts, flowers, agriculture, and an amazing array of esoterica, but certainly one of the leading varieties is the ethnic festival.[46] Although statistical documentation is meager, there is little doubt that such annual occasions have been multiplying greatly in recent years and that they are now widespread and diverse in character. Indeed, in some larger cities, such as Detroit, these are multiethnic spectaculars, with one group following another each succeeding week. Less ubiquitous, but still worthy of note, are the many new regional museums, frequently historical in tone, as well as the occasional ethnic museum and library.

Three attributes set the current ethnic and regional revivals apart from earlier modes of ethnic or regional existence. First, these are *self-conscious* phenomena among people who have been made aware, by one means or another, that they are rather special and ought to be doing something about that fact. In the past, members of particular regions or ethnic groups, living their lives spontaneously and without any narcissistic reflection, often were barely cognizant of their distinctiveness, sometimes only subconsciously so. Indeed, in the extreme case, they might be ashamed to acknowledge their identity. Second, these entities are no longer exclusive, private preserves of their members; they are meant to be touted and shared as widely as possible. In the immortal words of that classic subway ad, "You don't have to be Jewish to enjoy Levy's Rye Bread." Now we have transnationalism, or rather transregionalism and transethnicity, at the intimate scale. Consequently, and thirdly, commercial exploitation of ethnic and regional attractions has become rampant.

These essential features bear repetition: self-consciousness; a sharing of identities and cultural heirlooms with the world at large; and vigorous commercialization. However, I do not wish to oversimplify the situation. If there is all that touristic hype and schlock in portions of Lancaster County, some picturesque pockets of Teutonic culture in central Pennsylvania are still unmolested by outsiders. And not all of New England or New Mexico has been gussied up to ensnare the wallet of the passerby.

---

[45] Although necessarily quite incomplete, the most useful directories to American festivals of all types are Frances Shemanski, *A Guide to Fairs and Festivals in the United States* (Westport, Conn.: Greenwood Press, 1984) and Paul Wasserman and Edmond L. Applebaum, eds., *Festivals Sourcebook* (Detroit: Gale Reseach Company, 1984).

[46] My information on American ethnic festivals comes largely from an unpublished study by Catherine Harding, a former student at Pennsylvania State University.

Parallel with the revival and refashioning of ethnicity and regional-ism is the American's growing fascination with things of the past.[47] It materializes most poignantly in the search for "roots," in that new passion for family history, a genealogical craze that has infected a sizable fraction of the general public. Also at the personal level, in the most intimate of choices, is the fashionableness of "old-time" names for our offspring (among other interesting onomastic trends), especially the biblical vari-ety: Abigail, Adam, David, Deborah, Ethan, Jonathan, Joshua, Judith, Matthew, Michael, Sara, Timothy, but other venerable ones as well.[48] In selecting designs for new single-family homes during the 1970s and 1980s, Americans have decided in increasing numbers to go for recycled styles of the past, whether national or regional, as have the architects responsible for much postmodern commercial and office construction. And the bull market in antiques shows no signs of abating.[49] Within the public arena, the historical museum flourishes, as already noted, and so too the historical park, battlefield, neighborhood, site or building, and museum village. Historical pageants, centennial and similar festivities, reenactments of historical events, especially the bloodier varieties, and other costumed spectacles attract a growing clientele. The magnetism of a comforting past is undeniable.

The American Culture Area Revisited

For the geographer, the most revealing confirmation of the various trends touched upon in the preceding pages is visible in the character and dynamics of America's culture areas. Nothing that has come to my atten-tion during the past twenty years invalidates the tableau presented in Chapter 4 of this volume; but ongoing developments and further research and reflection enable one to supplement or refine the earlier observations in rather significant ways.

One of these ways is to draw a distinction between two varieties of the Traditional Region—between the Older and the Latter-day version (Table 5.1). While the former is the product of many centuries or even millennia of social and ecological evolution, a process of slow marination, so to speak, *in situ*, the Latter-day Traditional Region, which predomi-nates in North America and other neo-European settings as well, came into being rapidly. Indeed, in the case of the Mormon Culture Area,

---

[47] In *The Past Is a Foreign Country* (Cambridge: Cambridge University Press, 1985), David Lowenthal explores with impressive subtlety and insight the changing relationships and attitudes of Western societies, including the American, toward the historic past.

[48] I have been considerably startled lately to come across gravestones dating from the 1980s that were executed in an antique style, one that features the weeping willow motif. This was a practice that had become extinct almost everywhere by 1850.

[49] Quite a number of radio stations have exploited the nostalgia theme with some success, scheduling programs devoted solely to the popular music of earlier generations. A related development, but one not confined to the United States, began in the 1950s with a quickened interest in serious music of the Baroque Period and earlier. The more recent excitement over performance on authentic early instruments may be another sign of the times. The recycling of older fashions in women's clothing is still another phenomenon to be pondered.

Table 5.1.   A Typology of Culture Areas

Type	Dates	Location and Extent	Mode of Origin
Older Traditional Region	Prehistory to present	Nearly all premodern world; much of Old World today, plus scattered portions of Latin America	Interaction with local habitat and accessible aliens over extended periods; local inventions and cultural drift
Latter-day Traditional Region	Seventeenth century A.D. to present	Most neo-European lands, i.e., eastern North America, Siberia, Brazil, Argentina, Chile, Mexico, South Africa, Israel, Australia, New Zealand, inter alia	Scrambling of European, African, and other immigrants from varied source areas; interaction with new habitat and aboriginal populations
Voluntary Region	Late nineteenth century to present	Selected areas of varied magnitude in U.S. and other advanced countries	Spontaneous, selective migration of like-minded persons to places with desired qualities
Synthetic Culture Area	Late twentieth century	Scattered localities, many quite small, in U.S. and possibly, other postindustrial lands	Deliberate, self-conscious revival of traditional region or creation of make-believe places; may be ephemeral

genesis occurred virtually overnight. Moreover, in most cases we can document the origins and developmental scenario of such entities, though sometimes only by dint of intensive research.[50] Such relative historical transparency stands in strong contrast to the misty beginnings of nearly all of the older types.

The formation of the latter-day species has not ceased,[51] but is continuing by means of the time-honored processes of alien intrusion and mutual acculturation, as Joel Garreau has so convincingly argued for two tracts he calls "Mex-America" and "The Islands."[52] The former embraces a broad swath of the American Southwest along with the northern states of Mexico, while the latter is an archipelago including the West Indies and the Venezuelan coast but dominated by the recently hispanicized Miami metropolitan area. A comparable regional formation may be taking shape in northern Maine, Vermont, and New York with continuing infiltration by the Québecois. Because of the conditions of North American settlement history and demography, members of the newer generation of traditional culture areas tend to be larger than the ancestral type and to diverge less among themselves in terms of cultural and social characteristics. Their rate of change has also been markedly faster, especially in recent decades, because of unusually high, persistent levels of mobility among the native-born, substantial influxes of immigrants from sources other than the originating homelands, and the pervasive effects of modern modes of communication and transport.

Within the past two decades, American geographers have discovered and begun mapping and analyzing an important class of places that had somehow eluded earlier notice: Vernacular Regions. It is appropriate to discuss them here because, even though they may not be intrinsically cultural in nature, they can resemble, or even overlap, the genuine article. Terry Jordan has given us the definitive definition.

Perceptual or vernacular regions are those perceived to exist by their inhabitants and other members of the population at large. They exist as part of popular or folk culture. Rather than being the intellectual creation of the professional geographer, the vernacular region is the product of the spatial perception of average people. Rather than being based on carefully chosen, quantifiable criteria, such regions as are composites of the mental maps of the population.[53]

[50] Donald Meining has postulated a plausible scheme whereby the regions of the American West—with special attention to their demographic, circulation, political, and cultural aspects—have evolved over time and space in "American Wests: Preface to a Geographical Interpretation," *Annals of the Association of American Geographers*, 62, No. 2 (1972), 159–84.

[51] In *The Middle West: Its Meaning in American Culture* (Lawrence: University Press of Kansas, 1989), James R. Shortridge has chronicled the shifting core and boundaries of the region, at least as they are perceived by both insiders and outsiders. In another study of wavering regional identity, Carl Abbott, "Dimensions of Regional Change in Washington, D.C.," *American Historical Review*, 95, No. 5 (1990), 1367–93, we learn how that Janus-like metropolis has hovered between South and Northeast in its cultural affiliations over its 200-year career, but that recently it seems to be gradually tilting toward the latter.

[52] Joel Garreau, *The Nine Nations of North America* (Boston: The Houghton Mifflin Company, 1981), 167–244.

[53] Terry G. Jordan, "Perceptual Regions in Texas," *Geographical Review*, 68 (1978), 295–307.

One of the simpler, more obvious ways to ascertain the what and where of vernacular regions and, possibly, the cultural variety as well (notably the South and Midwest) is to interrogate the people in question.[54] That is exactly what a number of investigators have tried lately, and with useful results. Although the mail-back questionnaire has proved to be workable, at least when directed to the community elite, the more popular, foolproof method is to ply college students, a captive group, with the research instrument. Another more indirect methodology, but one that may be more effective in capturing the collective mental map of a population, was invented by sociologist John Shelton Reed and subsequently "plagiarized" by this author.[55] It involves plotting the incidence of certain terms having regional and/or cultural connotations that appear in the names of enterprises listed in telephone directories. The results are certainly interesting per se and may occasionally reinforce our faith in the reality of such cultural entities as the Southwest, Acadiana, the Bluegrass or the Illinoian segment of the Upper South colloquially known as Little Egypt. On the other hand, many of the vernacular regions are no more than locational perceptions—Eastern Shore, Panhandle, Gulf Coast, or Big Bend. Still others may be coinages invented by journalists, chambers of commerce, and other vested interests, items such as Metroplex in Texas, South Carolina's Grand Strand, North Carolina's Research Triangle, or California's Silicon Valley. Another category of vernacular region, such as the Bible Belt or Sun Belt, may be locationally vague while remaining prominent in our collective mental map. Awaiting investigation is the possibility that the more successful professional and college athletic terms may be generating yet another kind of regional identity within their fan sheds, whether or not such loyalties are translated verbally into viable regional terms.

Using another oblique approach, one without antecedent or sequel, at least to date, I have been able to discern the regional connotations of certain cultural and psychological attributes of the American population—to uncover still another layer of a geographical tapestry with a very thick pile. The study in question involved collecting statistics, at the relatively crude state level, on membership of truly voluntary associations and sales of special-interest magazines, a total of 163 such expressions of personal preferences ca. 1970.[56] The subsequent factor analysis disclosed the existence of several significant factors. It was reassuring to

[54]   The following are some of the more relevant studies: Joseph Brownell, "The Cultural Midwest," *Journal of Geography*, 59 (1960), 81–85; Ruth F. Hale, *A Map of Vernacular Regions in America*, Ph.D. dissertation (Minneapolis: University of Minnesota, Department of Geography, 1971); James K. Good, *A Perceptual Delimitation of Southern Indiana*, Professional Paper No. 8 (Terre Haute: Indiana State University, Department of Geography and Geology, 1976); Karl B. Raitz and Richard Ulack, "Appalachian Vernacular Regions," *Journal of Cultural Geography*, 2 (1981), 106–19; and Ari J. Lamme 2d and R. K. Oldakowski, "Vernacular Regions in Florida," *Southeastern Geographer*, 22 (1982), 100–109.
[55]   John Shelton Reed, "The Heart of Dixie: An Essay in Folk Geography," *Social Forces*, 54 (1976), 925–39; Wilbur Zelinsky, "North America's Vernacular Regions," *Annals of the Association of American Geographers*, 70 (1980), 1–16.
[56]   Wilbur Zelinsky, "Selfward Bound? Personal Preference Patterns and the Changing Map of American Society," *Economic Geography*, 50 (1974), 144–79.

find strong coincidence between personal preference patterns and both the Southern and Middle Western (cum Midland) regions, along with weaker, but persuasive, indications of linkages with the West and Southwest. But the single factor accounting for more variance than any other, the Urban-Migrant, is obviously more social or psychological than traditionally cultural; and its spatial expression resembles no other map I am aware of. The same statement applies to the other social/psychological factors, namely, Urban Sophistication, Migrant, and Sex and Romance, while the remaining two—the Latitudinal and Aquatic—are plainly associated with the physical habitat. Clearly much remains to be learned about the changing spatial structure of American society and its cultural components. We can no longer depend on the array of traditional cultural regions for more than a partial explanation of the territorial diversity of our population.

Returning to the more readily mappable culture areas of the United States, the next species to emerge following the maturation of the various traditional entities has been the Voluntary Region. The notion that such places, as defined and discussed elsewhere (pp. 134–139), evidently for the first time, are products of advanced societies seems to have won general acceptance. Over the past two decades the number and diversity of our Voluntary Regions have certainly increased. This is especially true for those based on retirement and recreational activity, some of which are seasonal or ephemeral in character. Another notable development has been the symbiotic merging or overlap of academic, research and development, and affluent retirement communities in and near a number of universities. The cases of Princeton, Stanford, the Research Triangle, Boulder, Colorado, and my own State College, Pennsylvania do not exhaust the inventory of examples. Still another subcategory, the Gay Enclave, has surfaced in recent times after having led a clandestine existence for some years.[57] San Francisco's Castro district may be the biggest and most flamboyant example, but virtually every one of our larger metropolises contains its counterpart.

The latest stage in the succession of modes of organizing cultural space in America is the still embryonic Synthetic Culture Area. Indeed, it is still so vaguely defined I must be tentative in claiming viability or generality. Although obviously akin to the Voluntary Region, which is usually spontaneous in origin, the newer breed departs from it by virtue of being quite self-conscious and deliberate, by its total artificiality. The majority of these creations—the handiwork of local enthusiasts, business interests, and governmental agencies—represent efforts to resuscitate or reinvigorate real or imagined localities of the past; but a minority, such as some of the more imaginative Edge Cities, are completely novel in concept. Certain gentrifying urban neighborhoods fall into this realm of make-believe, along with a number of carefully contrived and controlled shopping malls, suburban residential developments, or the likes of Seaside, Florida.

Such ersatz confections are not entirely new; what is newsworthy

---

[57] See note 22

about them today is their number and diversity. As far back as the 1890s, that Gothic oasis known as the University of Chicago appeared on the Midway in Chicago's South Side. The lived-in Historic Williamsburg, along with more recent historic villages and theme parks, might also bear the label of Synthetic Culture Area. For some time also we have had the spectacle of those pseudo-Mexican tourist districts in the towns just south of the border, synthetic micro-regions so closely tailored to the fantasies of their Yanqui patrons. At the miniature scale, the trend is clear enough in many individual dwellings, restaurants, and other shops. The connections with postmodernity and Landscape Three and to the Ethnic and Regional Revivals are quite unavoidable.

But, in a larger sense, this penchant for the making or remaking of culturally soothing habitats represents the American Dream, or experiment, come full circle. For, in a recognizable way, such synthesizing has been going on in the United States for more than two hundred years in that grandest, most fully consummated of artificial culture areas—the modern nation-state. Insofar as the United States (and France, Italy, Israel, or Mexico for that matter) has become a distinctive and relatively homogeneous place in a cultural sense—and it most certainly has—it is largely through the workings of a powerful, determined central state apparatus and related business interests with their leverage over so many phases of our social and economic activities. We have in America an ideal example of the synthetic region at the macro scale, a community that began as a novel sort of nation with its unique brand of peoplehood but later turned into a full-fledged nation-state because it was fashioned essentially *de novo* without the ancient roots, the real or imagined age-old traditions, that the budding European nation-states were able to bank on so effectively.

The Question of Convergence

Up to now I have carefully avoided confronting a question as profound, urgent, and, indeed, troubling as any the social scientist is likely to encounter today (although brushed over lightly at an earlier point [pp. 87–88]): Is our human world being ground down into a uniform place with no meaningful distinctions any longer from one locality to another? Is space-time convergence making a mockery of the age-old peculiarities of nations and regions? Or if not for the world as a whole, at least within such vanguard countries as the United States? There can be little doubt about the impact of the pulverizing forces of modernization, as noted repeatedly in the previous pages—the instantaneous saturation of vast populations with all manner of information via telecommunications; all those wizardly new technologies for mass-producing goods, services, and attitudes and for transporting people and things far, wide, and swiftly; mechanisms that cannot help but lessen differences among places. Indeed, the conventional, taken-for-granted wisdom among laypersons and most scholars alike is that we now inhabit a drastically shrunken world in which place-to-place diversity is vanishing. Some soothsayers have even proclaimed an end to both history and geography, thereby heralding the dawn of some sort of event-less, featureless nirvana.

The truth is rather more complicated and interesting. First of all, there has been remarkably little substantive research on whether cultural convergence has, in truth, been occurring within the American nation-state or the world at large, and, if so, how or to what extent. Such evidence as we do have suggests a dialectical situation, of countervailing forces perpetually at odds and yielding new, usually unexpected social and geographic complexities. This is clearly what is happening within the economic realm. Innovations in systems of finance, management, production, and distribution have meant many new areas of specialization and the abandonment of obsolete centers. Thus the spatial divisions and concentrations of labor and investment are continuing while falling into new patterns in many branches of manufacturing and in the tertiary and quaternary industries. In parallel fashion, if we take the trouble to inspect the actuality of the American world about us, we find not only the pervasive commonalities (which, in various forms, have always characterized this part of the world) but also fresh local pockets of social and cultural individuality. They include all those many Voluntary Regions that make for a crazy quilt of diversity and also those feigned space-warps and time-warps I have called Synthetic Regions.[58]

But even if we dismiss such peculiar spots as aberrations from the norm, the large residual sociocultural map fails to conform with the hypothesis of an impending human peneplain. In a provocative, if jarringly journalistic, account, Michael Weiss has reviewed the results of an intensive analysis of both small-area Census data and information concerning consumer behavior at the Zip code area level assembled from a variety of unofficial sources ca. 1980.[59] What emerges from this massive data bank is a constellation of some forty social, economic, demographic, and, implicitly, cultural localities or "clusters." Each cluster is an archipelago of widely scattered but similar localities, not a contiguous block of territory. This is not a portrait of a homogenized, pureed America but rather a country where people have been sorting themselves into compatible niches, where notably sharp distinctions prevail within individual cities and stretches of countryside—and, contrary to popular myth, within suburbia as well. Even more directly to the point at issue here, Weiss introduces some evidence strongly suggesting deepening disparities among these ever-changing clusters during the decade 1970 to 1980.

The uneasy conclusion, then, is that there is no pat answer to the question that unnerves so many of us as to whether the United States and the entire world are melting into a single unified cultural blob—there is only the ambivalent yes and no. My own hunch is that we can never outgrow or dissolve regional specialness, whether within North America or at the global scale. The ultimate reality is that the nature of each and every region, of whatever type, has forever been changing and, in all

---

[58] The theme of a convoluted interplay between global and localistic developments is one that recurs throughout Anthony Giddens' splendid meditation on *The Consequences of Modernity* (Stanford: Stanford University Press, 1990), but perhaps most explicitly in geographic terms in pp. 140–143.

[59] Michael J. Weiss, *The Clustering of America* (New York: Harper & Row, Publishers, 1988).

likelihood, will insist on evolving forever and in multiple directions. We can serve our material and intellectual interests well by keeping a sharp lookout for these changes.

Envoi

How is one to present a tight and tidy summation of anything so untidy and complex as the changing cultural geography of the United States in the very late twentieth century? The prudent, but impractical, strategy would be to wait one hundred or two hundred years until the dust and din have settled. But since all of us are impatient, this must be an interim report, one observer's best guess as to which basic processes and trends are truly critical and most worth watching.

I find just a handful of such central themes, developments that do not have autonomous lives of their own but rather impinge upon, and interact, with each other. As fundamental a fact as any is one that is transgeographic: that America is approaching a turning point, some sort of crisis of identity, after more than three centuries of wild, headlong expansion and flexing of muscle. Limitations—economic, ecological, and institutional— are becoming all too visible in both the domestic and global arenas, so that this is a country no longer as dominant and buoyant as in the recent past. A restructuring of our collective *mentalité* seems to be progressing, one that places us decidedly in the company of other postindustrial nations.

But, coming down to earth from such cosmic considerations, there are the mappable actualities that geographers are primed to wrestle with. It is heartening to discover that the emerging new order of things is distinctly legible in the volatile landscape features and settlement morphology of the United States as well as in those less visible, but place- and time-specific acts and attributes that also fall within our domain. For this author, the single greatest revelation has been the realization of the dialectic nature of our changing scene, "the juxtaposition or interaction of conflicting forces, ideas, etc.," to quote the lexical definition. Never has it been more evident that social history and evolving human geographies do not slide along straight regular grooves. Thus we can discern that, working at cross-purposes against the all too obvious process of convergence and homogenization, are innumerable local foci of resistance, all those spatially intimate and personal assertions of individuality and specialness. Given their timing and self-awareness, the regional and ethnic revivals would seem to qualify as counterattacks against centralizing pressures toward standardization at the national and international levels. And, as a rebellion against the tarnished vistas of the future, we have that most emphatic reversion to the solace of past worlds, everywhere that one might look a nostalgia for what should or might have been.

The latest phase of America's cultural geography is characterized not only by the interaction and separate trajectories of conflicting forces but also by a degree of transnational interdependence with societies far and near quite unlike anything previously known. Transnationalization is as significant a development in our cultural existence as any other touched upon in this volume. In this massive long-distance sharing of so many elements of folk, popular, and elite cultures, the American community has almost certainly been more active as transmitter and receiver than any other. However shaky our economic prowess or however transitory American military primacy may prove to be, it seems unlikely that any other country can soon usurp our role as the pivotal actor in the great game of transnationalizing world culture.

The material treated in this chapter serves one ultimate purpose: to remind us of the mutability of all things human. Although intellectually we may know better, it is all too easy to assume the fixity of certain institutions, ideas, or other abstractions. I trust that the review of the history of American nationalism and statism illustrates how time and circumstance can alter even the most seemingly solid of values. And I close, finally (the loveliest word in the English language), with the prayer that the foregoing observations will make some sort of sense to any readers I may have in the twenty-first century.

Selected References

The published literature available to the student of American cultural geography is vast in quantity (and indeed much too enormous for this bookworm of an author ever to master), immensely broad and diversified as to topic, but also uneven in quality and depth of coverage. Consequently, the following list is selective, and is intended to suggest to the serious reader only the more notable works I am aware of in some seventeen categories, along with one section for miscellany. The zealous student will discover that nearly all the recommended publications contain numerous bibliographic cues for more detailed exploration. The absence of annotation indicates an item of substantial merit whose title adequately describes its contents. (Although I am including only printed matter, I am well aware of the availability of some excellent documentary films and videocassettes. Maybe next time.) Happy perusing!

1. *General Bibliography and Reference Works*

FRANK B. FREIDEL, ed., *Harvard Guide to American History*, rev. ed., 2 vols. Cambridge, Mass.: Harvard University Press, 1974. A massive bibliography covering not only traditional historical items but also virtually every aspect of American life and civilization.

CHAUNCY D. HARRIS, *Bibliography of Geography. Part 2: Regional. Volume 1. The United States of America* (Research Paper No. 206). Chicago: University of Chicago, Department of Geography, 1984. One of many bibliographic tours de force by the amazing Professor Harris. This exemplary volume contains 974 annotated entries dealing with every imaginable aspect of American geography.

———— ET AL., *A Geographical Bibliography for American Libraries*. Washington, D.C.: Association of American Geographers and National Geographic Society, 1985. Amidst the 2,903 annotated entries in this carefully edited volume covering the entire field of geography are many items bearing, directly or indirectly, on America's cultural geography. Cited hereinafter as GBAL.

HUGH C. PRINCE, "Three Realms of Historical Geography: Real, Imagined and Abstract Words of the Past," *Progress in Geography*, 3 (1971), 4–86. A major review article, with the bulk of the references directed to the North American scene.

U.S. BUREAU OF THE CENSUS, *Statistical Abstract of the United States*. Washington, D.C.: Government Printing Office, 1879–. Annual. "The one publication that is absolutely essential for anyone concerned with the U.S. In addition to the 1500 tables furnishing data on virtually every conceivable subject [including the cultural], this volume is a guide to sources." (GBAL)

U.S. LIBRARY OF CONGRESS, *A Guide to the Study of the United States of America: Representative Books Reflecting the Development of American Life and Thought*. Washington, D.C.: Government Printing Office, 1960. *Supplement, 1956–1965*. Washington, D.C.: Library of Congress, 1976. "A selection of 6487 titles [in the 1960 volume], liberally annotated, that afford an excellent introduction to every form of inquiry concerning the United States. Of special value are chapters 6 and 12, dealing with geography and local history. . . . This supplement provides 2943 additional titles published over the span of one decade. The arrangement is by subject." (GBAL)

2. *Periodicals*

American Demographics. Ithaca, N.Y.: 1979–. Monthly. Although predominantly concerned with consumer behavior and matters demographic and strongly slanted toward the bottom-line interests of its marketing research and sales promotion clientele, this magazine does carry occasional worthwhile articles and shorter pieces that document the changing cultural-geographic scene.

Journal of Cultural Geography. Bowling Green, Ohio: Popular Culture Press, 1980–. Semiannual. If one is willing to overlook a laissez-faire policy in accepting submissions, poor editing, and the wretched physical production so characteristic of the periodicals and books gushing out of the publishing mill at Bowling Green State University, there are some real gems to be found amidst the dross of this journal.

Journal of Historical Geography. London and New York: Academic Press, 1975–. "Scholarly articles on historical geography with a wide range of areas and periods studied. Debates. Review articles. Extensive reviews. Short notices." (GBAL)

Journal of Popular Culture. Bowling Green, Ohio: Popular Culture Press, 1967–. Quarterly. See comments on *Journal of Cultural Geography*.

Journal of Regional Cultures. Bowling Green, Ohio: Popular Culture Press, 1981–. Semiannual. See comments on *Journal of Cultural Geography*.

Landscape. Berkeley, Calif.: 1951–. (suspended 1971–1974) Issued three times a year. This remarkable journal, founded by J.B. Jackson and edited and published by him out of Santa Fe, N.M., for many years, has been the single most influential factor in the flowering of serious landscape scholarship in America. Indispensable reading for the cultural geographer and landscape ogler—and beautifully designed and printed.

Material Culture: Journal of the Pioneer America Society. Pioneer America Society: 1969– (published as *Pioneer America*, 1969–1983). Issued three times a year.

North American Culture. Stillwater, Okla.: North American Culture Society, 1984–. Semiannual. Scholarly articles on all aspects of the cultural and folk geography of the United States.

3. *Cultural Geography—General Works*

SAMUEL N. DICKEN and FORREST R. PITTS, *Introduction to Cultural Geography*. Waltham, Mass.: Xerox College Publishing, 1971. A revision of the original 1963 edition and one that adopts the topical approach. Several of the chapters are innovative in coverage.

PETER JACKSON, *Maps of Meaning: An Introduction to Cultural Geography*. London: Unwin Hyman, 1989. Although it includes a historical review of work in the area, this is basically a programmatic statement urging closure between cultural and social geography and embodying much of the agenda of today's politically and socially concerned scholars. Most of the substantive material is British, but there are ample references to the American scene.

TERRY G. JORDAN and LESTER ROWNTREE, *The Human Mosaic: A Thematic Introduction to Cultural Geography*, 5th ed. New York: Harper & Row, Publishers, 1990. Although this highly successful introductory text is global in reach, there is particular emphasis on North American items in many of its pages. Treatments of folk and popular geography appear here for the first time in the textbook literature.

ALFRED L. KROEBER, *The Nature of Culture*. Chicago: University of Chicago Press, 1952. As definitive a statement as one could have hoped for a generation ago on a profoundly difficult question by one of the foremost anthropological thinkers.

DAVID LEY, *A Social Geography of the City*. New York: Harper & Row, Publishers, 1983. Given the fact that so much of our human geography has become urban, this sensitive and innovative approach to the city richly merits inclusion in this corner of our reference guide.

JOSEPH E. SPENCER and WILLIAM L. THOMAS, JR., *Cultural Geography: An Evolutionary Introduction to Our Humanized Earth*. New York: John Wiley & Sons, 1969. A richly rewarding and erudite monograph, disguised as a textbook, that is rich in theory, fact, and exceptional maps and photos. The book employs a genetic, or evolutionary, approach to cultural process and form.

PHILIP L. WAGNER, *The Human Use of the Earth*. New York: Free Press, 1960. Discusses the elements of cultural geography and proposes a classification of these elements and processes. A broad synthesis of the relevant findings of several disciplines.

———— AND MARVIN W. MIKESELL, eds. *Readings in Cultural Geography*. Chicago: University of Chicago Press, 1962. Even more valuable than the judicious selections from scholars in human geography and related disciplines are the editors' deeply probing Introduction and the prefatory comments accompanying each section.

4. Atlases

LESTER J. CAPPON, ed., *Atlas of Early American History: The Revolutionary Era, 1760–1790*. Princeton, N.J.: Princeton University Press, 1976. A full-color production that is impressive in terms of both scholarship and cartography. The volume may reflect the historian's mind-set more than the geographer's, and it contains ample coverage of the predictable economic, political, and military topics, but there is also much documentation of matters social and cultural.

WILBUR E. GARRETT, ed., *Historical Atlas of the United States*. Washington, D.C.: National Geographic Society, 1988. Despite having been generated quite hastily and displaying an unseemly high ratio of pictorial matter to map and text, this large, visually appealing volume is the worthiest successor to date to the mighty Paullin-Wright atlas. Laudably catholic in probing all significant aspects of American life and richly laden with bibliographic citations.

R. COLE HARRIS, ed., *Historical Atlas of Canada. Volume I: From the Beginning to 1800*. Toronto: University of Toronto Press, 1987. This publication, the first of a three-volume set, may well be the finest historical atlas ever produced anywhere in terms of both scholarship and cartographic sophistication. It appears here because the human geography of the United States and the future Canada were so inextricably intertwined during their formative years.

THEODORE R. MILLER, *Graphic History of the Americas*. New York: John Wiley & Sons, 1969. The 61 black-and-white plates in this atlas provide much useful, well-arranged information on the historical geography of the Western Hemisphere.

CHARLES O. PAULLIN and JOHN K. WRIGHT, *Atlas of the Historical Geography of the United States*. Washington and New York: Carnegie Institution and the American

Geographical Society, 1932. The American geographer exiled to a desert island with a baggage limit of a single book should give this item serious consideration despite its antiquity. One of the truly mighty monuments of American historical and geographic scholarship. Most of the plates, dealing with a broad range of physical, historical, social, political, and economic topics, were painstakingly compiled from multiple sources. Detailed notes on source materials.

READER'S DIGEST ASSOCIATION, *These United States: Our Nation's Geography, History and People.* Pleasantville, N.Y.: Reader's Digest, 1968. The best approximation of a national atlas until the genuine article arrived two years later. Imaginative design and useful coverage of a broad range of topics at both national and regional scale.

JOHN F. ROONEY, JR., WILBUR ZELINSKY, and DEAN R. LOUDER, eds., *This Remarkable Continent: An Atlas of United States and Canadian Society and Culture.* College Station: Texas A&M University Press, 1982. An anthology of 387 previously published and unpublished black-and-white maps grouped into 13 sections, each prefaced by a short essay. Includes material on such relatively neglected topics as foodways, music and dance, sports and games, and place perception. Sui generis.

BARBARA GIMLA SHORTRIDGE, *Atlas of American Women.* New York: Macmillan, 1987. The combination of 128 state-level choropleth maps on all currently mappable topics with ample text and tabular material provides us with our first comprehensive look at the geography of American women. Most of the data are for 1980 or thereabouts, but there are some forays into the past.

U.S. DEPARTMENT OF THE INTERIOR, GEOLOGICAL SURVEY, *The National Atlas of the United States of America.* Washington, D.C.: Government Printing Office, 1970. Although there are some distressing lacunae (as, for example, in such areas as ethnicity and language), the United States at long last covered itself with cartographic glory in this attractive, but out of print, volume. The plates and marginal comments on population, settlement, and social structure are quite handsomely executed. Will the government of one of the world's richest countries ever come up with funds for the much-needed sequel?

5. *General Treatments of the United States—and Its Historical Geography*

CONRAD M. ARENSBERG and SOLON T. KIMBALL, *Culture and Community.* New York: Harcourt Brace Jovanovich, 1965. A collection of their earlier papers by two social anthropologists dealing with the theory and substance of American community studies. Although rather fuzzy at times, this is probably the best statement on the topic.

RALPH H. BROWN, *Mirror for Americans: Likeness of the Eastern Seaboard, 1810.* New York: American Geographical Society, 1943. A veritable tour de force of the geographical imagination. The regional geography of the United States in 1810 as it might have been written by a contemporary geographer.

J. C. FURNAS, *The Americans: A Social History of the United States, 1587–1914.* New York: G.P. Putnam's Sons, 1969. This vast (1,015 pages), sprawling, invertebrate *omnium gatherum* of a book treats just about every imaginable item in American social history and material culture while studiously avoiding matters political or military. Its avoirdupois is relieved by the author's wry wit.

PAUL FUSSELL, *Caste Marks: Style and Status in the U.S.A.* London: Heinemann, 1984. The persistence of a well-entrenched class structure in America is one of those dirty secrets few of us are willing to acknowledge in a land where almost everyone claims membership in the middle class. Fussell has no such inhibitions, and he explores class, status, and the outward signs thereof with much panache and wit, though too often with a supercilious sneer. Immensely readable.

PETER KALM, *Travels in North America: the English Version of 1770,* rev. and ed. by Adolph E. Benson, 2 vols. New York: Dover Publications, 1966. Perhaps the most notable and informative of the firsthand accounts of colonial British America, especially the middle and northern colonies, written by a Swedish natural scientist with a keen eye and catholic curiosity.

MAX LERNER, *America as a Civilization: Life and Thought in the United States Today.* New York: Simon & Schuster, 1957. Encyclopedic observations and reflections on the grand theme of the nature and meaning of the American experience by an able scholar and journalist.

FERDINAND LUNDBERG, *The Rich and the Super-Rich: A Study in the Power of Money.* New York: Lyle Stuart, 1968. A massive, acidulous, and important treatise on the social and political implications of the concentration of property and power, with some sharp geographic insights. Muckraking in the grand manner, with ample documentation.

D. W. MEINIG, *The Shaping of America: A Geographical Perspective on 500 Years of History. Volume I: Atlantic America, 1492–1800.* New Haven, Conn.: Yale University Press, 1986. I am hardly alone in regarding this first installment of a probable tetralogy as the single finest book ever produced by any American geographer on any subject. *In any case, if you read no other book in this entire list, do read this!* Employing eloquent prose and imaginative graphics, Donald Meinig gives us an original, persuasive conspectus of the historical geography of the Atlantic World as the Europeanization of the Americas took shape, with deep insights into the social, cultural, and spatial processes at work in the confrontations among varied immigrant and indigenous groups on this continent.

ROBERT D. MITCHELL and PAUL A. GROVES, eds., *North America: The Historical Geography of a Changing Continent.* London: Hutchinson, 1987. This collection of 18 chronologically ordered essays by a set of ideal authors is clearly the best currently available treatment of the evolution of the United States and Canada over time from the days of early exploration to the late twentieth century. Highly recommended.

ANDRE SIEGFRIED, *America at Mid-Century.* New York: Harcourt Brace Jovanovich, 1955. A refreshing, deeply perceptive, highly readable view of the nature of the American land and people by a widely travelled French geographer. Thematically organized, with discussions of ethnic and religious phenomena.

GEORGE R. STEWART, *U.S. 40: Cross Section of the United States of America.* Boston: Houghton Mifflin Company, 1953. A magnificent interpretation of the American landscape that is a mixture of travelogue, photograph, caption, and essay. Much more than just a book about a road.

W. LLOYD WARNER, *American Life: Dream and Reality.* Chicago: University of Chicago Press, 1953. Ruminations by a distinguished sociologist based in large part on his intensive studies of a small New England city. Essentially an anthropological analysis of some basic attributes of the American social structure and value system.

J. WREFORD WATSON, *North America, Its Countries and Regions,* rev. ed. New York: Praeger, 1967. Among college texts dealing with this continent, Watson's is arguably the most literate and intellectually nourishing, and is the only one according culture and history more than a nodding glance. A useful basic reference designed for the British public.

———, *Social Geography of the United States.* London and New York: Longman, 1979. Essentially a group of essays on the social and ethnic problems of the country viewed in geographic perspective. Commendable.

MICHAEL J. WEISS, *The Clustering of America.* New York: Harper & Row, Publishers, 1988. A breezy account of the social geography of the United States as a mosaic of 40 neighborhood types or clusters as ascertained through consumer behavior as well as the usual socioeconomic criteria. Although it sounds all too often like a book-length blurb for the Claritas Corporation, this item has the virtue of originality and some startling revelations about the spatial restructuring of our population. The data tend to support some of my speculations about territorial self-sorting in Chapter 4.

THOMAS C. WHEELER, ed., *A Vanishing America; The Life and Times of the Small Town.* New York: Holt, Rinehart & Winston, 1964. Twelve beautifully written essays, each by a gifted author dealing with a place he knows and cherishes. Wallace Stegner's introduction is memorable.

WILBUR ZELINSKY, "Selfward Bound? Personal Preference Patterns and the Changing Map of American Society." *Economic Geography*, 50 (1974), 144–79. A state-level study, exploiting factor analysis and based on data on readership/membership for 163 special-interest magazines and voluntary organizations, 1970–1971. Some of the 10 patterns that emerge are familiar; others are new and startling.

————. *Nation into State: The Shifting Symbolic Foundations of American Nationalism.* Chapel Hill: University of North Carolina Press, 1988. An exploration—the first full-scale effort of its kind for any modern nation-state—of the interplay between symbols in all their many forms and the development and transformation of nationalism. Essentially historical in structure, but with some useful geographical sidelights.

6. National Character

JOHN GILLIN, "National and Regional Cultural Values in the United States," *Social Forces*, 34 (1955, December), 107–13. Every word counts in this pithy essay.

MICHAEL A. GOLDBERG and JOHN MERCER, *The Myth of the North American City.* Vancouver: University of British Columbia Press, 1986. No, this item is not out of place here. In a superb comparative study, the authors attribute the obvious sharp differences in the ways Canadian and American cities look and function to contrasts in the social and political axioms of the two countries and thus, implicitly, in their national characters. 'Tis a pity that good cross-national analysis is such a rare commodity.

FRANCIS L. K. HSU, *American and Chinese: Two Ways of Life.* New York: Henry Schumann, 1963. A truly superb analysis by a "marginal man," an anthropologist deeply rooted in both Chinese and Euro-American life. The theme throughout is the dominant role of mutual dependence within the primary group among the Chinese and its absence among Americans.

HOWARD MUMFORD JONES, *O Strange New World: American Culture; The Formative Years.* New York: Viking, 1964. This account by a cultural historian of the shaping of our national language, laws, religion, ideals, education, literature, and arts through the interaction of Old World and New World forces is original, insightful, and readable.

MICHAEL MCGIFFERT, ed., *The Character of Americans; A Book of Readings.* Homewood, Ill.: Dorsey Press, 1965. An excellent selection of essays, old and new, on the peculiarities of *Homo Americanus.* Opulent bibliography.

PERRY MILLER, *Nature's Nation.* Cambridge, Mass.: Harvard University Press, 1967. A posthumous collection of worthy essays on a variety of American themes by the historian who labored so diligently and productively on the intellectual evolution of New England.

HUGO MUNSTERBERG, *The Americans.* New York: McClure, Phillips & Co., 1905. A compendious interpretation of the American scene for the German public by a German psychologist teaching at Harvard. Well below the level of the greatest items in this genre, but still interesting for its facts and biases.

GEORGE JEAN NATHAN and H. L. MENCKEN, *The American Credo: A Contribution toward the Interpretation of the National Mind.* New York: Knopf, 1920. Superficially this appears to be nothing more than a string of epigrams by the two reigning wits of the 1920s. In actuality, a great deal of shrewd observation and wisdom lies beneath the glittering surface.

VANCE PACKARD, *A Nation of Strangers.* New York: D. McKay, 1972. A lively, semipopular account of restless Americans and their phenomenal territorial mobility.

GEORGE PIERSON, *The Moving American.* New York: Knopf, 1973. Another monograph that, like Vance Packard's, tells us what is known about the itching-foot malady of the American and what it means, but does so in a more academic fashion.

DAVID M. POTTER, *People of Plenty.* Chicago: University of Chicago Press, 1954. As penetrating a discussion of the vexed issue of national character as one could hope to find, and also a persuasive argument for the importance of material abundance and a related social fluidity in shaping the American version thereof.

ALEXIS DE TOQUEVILLE, *Democracy in America,* ed. by Henry Steele Commager. New York: Oxford University Press, 1947. The scholarly consensus is correct. This remarkable, enduring work (written in the 1830s) is the most perceptive commentary yet on a distinctively new nation.

7. Historical Geography of Settlement and the Economy

JAMES AXTELL, *The Invasion Within: The Contest of Cultures in Colonial North America.* New York: Oxford University Press, 1985. One of the more heartwarming developments on the scholarly scene in recent years has been the burgeoning of studies on European-aboriginal relationships, frequently in a revisionist vein. Axtell's account is one of the more sensitive and informative.

BERNARD BAILYN, *Voyagers to the West: A Passage in the Peopling of America on the Eve of the Revolution.* New York: Knopf, 1986. An extraordinary production in every respect. The availability of a unique set of emigration data created by the British Board of Trade for a brief period during the 1770s has enabled this distinguished historian to document in great detail just where the emigrants haled from, what sorts of folks they were, where they embarked, and where they wound up in North America. A good many personal vignettes enliven the account, one that stands alone in its social and geographical specificity.

RAY A. BILLINGTON, *Westward Expansion: A History of the American Frontier.* New York: Macmillan, 1967.

RALPH H. BROWN, *Historical Geography of the United States.* New York: Harcourt, Brace, 1948. The first effort of its kind, a scholarly, well-written narrative and geographic analysis of the settlement and economic development of the major regions of the United States until about 1870.

ALFRED W. CROSBY, JR., *The Columbian Exchange: Biological and Cultural Consequences of 1492.* Westport, Conn.: Greenwood Press, 1972. The emphasis is heavily on the biotic exchange, i.e., wild and domesticated plants and animals, disease, and human genetic material. The best item of its sort to date.

DAVID HACKETT FISCHER, *Albion's Seed: Four British Folkways in America.* New York & Oxford: Oxford University Press, 1989. This bulky, erudite study, the first installment of a projected multivolume cultural history of the United States, carries the Doctrine of First Effective Settlement to an extreme, by insisting that *all* of American cultural geography and political history can be explained by the attributes of early settlers from four particular sections of Great Britain and their transplantation into four colonial culture hearths. However flawed the central thesis may be, the book presents a wealth of useful information on the communities in question.

HERMAN R. FRIIS, *A Series of Population Maps of the Colonies and the United States, 1625–1790.* (Mimeographed Publication No. 3.), New York: American Geographical Society, 1940; rev. ed. 1968. Minutely detailed maps, accompanied by extensive documentation. An indispensable source for the study of early frontiers and of population and settlement patterns.

JOHN HERBERS, *The New Heartland: America's Flight Beyond the Suburbs and How It Is Changing Our Future.* New York: Times Books, 1986. A thoughtful treatment of the exurbanization of the American population and its implications.

KENNETH T. JACKSON, *Crabgrass Frontier: The Suburbanization of the United States.* New York: Oxford University Press, 1985. A detailed examination of the physical and social evolution of American cities and suburbs over more than 200 years, and the causes and consequences thereof. Considered the most nearly definitive such effort to date.

HILDEGARD BINDER JOHNSON, *Order upon the Land: The U.S. Rectangular Land Survey and the Upper Mississippi Country.* New York: Oxford University Press, 1976.

Although regionally focused, this fine treatment of the history and implications of this dominant mode of land survey and allocation is relevant to virtually all of trans-Appalachian or postcolonial America.

TERRY G. JORDAN, "Preadaptation and European Colonization in Rural North America," *Annals of the Association of American Geographers*, 79 (1989), 489–500. This presidential address offers a hypothesis—one central to this major cultural geographer's research—on the cultural ecology of the American settlement process that is thought-provoking and possibly persuasive.

——— and MATTI KAUPS, *The American Backwoods Frontier: An Ethnic and Ecological Interpretation*. Baltimore: The Johns Hopkins University Press, 1989. A revisionist approach to the genesis and spread of material culture on the American frontier. Some readers will be persuaded by the voluminous field and documentary evidence from Scandinavia and the United States of the soundness of the authors' thesis—that this transitory cultural complex originated in the lower Delaware Valley with the arrival of Finnish and Swedish backwoodsmen and some borrowings from the local aborigines, and then diffused widely. The debate may not be finished.

FRANCIS J. MARSCHNER, *Land Use and Its Patterns in the U.S.* (U.S.D.A., Agriculture Handbook No. 153) Washington, D.C.: Government Printing Office, 1959. "The first 100 pages of this classic monograph treat land use patterns and land survey from the perspective of their historical development. The remainder of the volume is a choice selection of air photos showing the patterns and the general appearance of various portions of rural America." (GBAL)

DOUGLAS R. MCMANIS, *Historical Geography of the United States: A Bibliography (including Alaska and Hawaii)*. Ypsilanti: Eastern Michigan University, Division of Field Services, 1965. An extensive compilation of titles arranged by region and topic, but without annotation. Concentrates mainly on land settlement and the economy.

RANDALL D. SALE and EDWIN D. KARN, *American Expansion: A Book of Maps*. Homewood, Ill.: Dorsey Press, 1962. Twelve neatly executed maps, accompanied by brief commentary, covering the period 1790–1900. The identification of land offices is a rare and welcome feature.

NORMAN J. W. THROWER, *Original Survey and Land Subdivision: A Comparative Study of the Form and Effect of Contrasting Cadastral Surveys*. Chicago: Rand McNally, 1966. A detailed analysis of land disposal in Ohio and the first implementation anywhere in America of the rectangular survey system.

FREDERICK JACKSON TURNER, *The Frontier in American History*. Melbourne, Fla.: Krieger, 1953. A collection of 13 major essays by Turner. His contention that the frontier experience was central to the development of American life and institutions has had a major and lasting impact on scholarly thought and controversy.

DAVID WARD, *Cities and Immigrants: A Geography of Change in Nineteenth-Century America*. New York: Oxford University Press, 1971.

MICHAEL WILLIAMS, *Americans and Their Forests: A Historical Geography*. New York: Cambridge University Press, 1989. The definitive monograph, one that covers all aspects of American forests and their exploitation past and present.

8. Attitudes toward the Land; Human Ecology

PETER BLAKE, *God's Own Junkyard: The Planned Deterioration of America's Landscape*. New York: Holt, Rinehart & Winston, 1963. Written "not in anger, but in fury." This combination of photograph, quotation, and brief text is a powerful polemic against those who have raped the environment for private gain. Intemperate but instructive.

WOLFGANG BORN, *American Landscape Painting*. New Haven, Conn.: Yale University Press, 1948. One of the better surveys of a subject rich in potential interest for the cultural geographer.

194   Selected References

WILSON O. CLOUGH, *The Necessary Earth: Nature and Solitude in American Literature*. Austin: University of Texas Press, 1964. A volume of modest achievement. Through sheer persistence, Clough does make his point about the centrality of wild nature as a literary, and thus basic cultural, theme in American experience.

WILLIAM CRONON, *Changes in the Land: Indians, Colonists, and the Ecology of New England*. New York: Hill & Wang, 1983. An instant classic. This superlative, subtle, non-ethnocentric monograph deftly describes and interprets the ways in which varied aborigines and European settlers interacted over time with a habitat that varies widely within the region.

ARTHUR E. EKIRCH, *Man and Nature in America*. New York: Columbia University Press, 1963.

HANS HUTH, *Nature and the American: Three Centuries of Changing Attitudes*. Berkeley: University of California Press, 1957. A thoughtful, useful survey of the topic, crammed with much detail concerning artists, travelers, conservationists, vacationers, and others.

ANNETTE KOLODNY, *The Lay of the Land: Metaphor as Experience and History in American Life and Letters*. Chapel Hill: University of North Carolina Press, 1975. The double entendre in the main title gives the plot away in this widely noted feminist tract arguing predominantly macho attitudes among American writers and doers toward their land, visualized here as the ravished female body. Overdone and flawed, but unforgettable.

DAVID LOWENTHAL, "The American Scene," *Geographical Review* 58 (1968), 61–88. Delves into fundamental issues. Must reading.

LEO MARX, *The Machine in the Garden: Technology and the Pastoral Ideal in America*. New York: Oxford University Press, 1964. A most influential volume—indeed, an acknowledged classic. Largely on the basis of literary evidence, the zone of tension among the forces of technology, the wilderness ideal, and the pastoral vision in the American mind is charted most expressively.

IAN NAIRN, *The American Landscape: A Critical View*. New York: Random House, 1965. A provocative melange of photo and prose by a British architect that is more than a simple jeremiad. The central argument is that the basic flaw in the visible landscape (and inferentially in American life as a whole?) is the weakness or absence of genuine relationship and identity—an outgrowth of the essential privatism of Americans.

RODERICK NASH, *Wilderness and the American Mind*, 3rd ed. New Haven, Conn.: Yale University Press, 1982. A sensitive, thoughtful, erudite chronicle of American (and European) attitudes toward wilderness in all their ambiguity. Extensive bibliography.

FORREST R. PITTS, "Japanese and American World-Views and Their Landscapes," in *Proceedings of IGU Regional Conference in Japan 1957* (pp. 447–59). Tokyo: 1957. Comparative cultural geography as it should be written. Elegant.

CHARLES L. SANFORD, *The Quest for Paradise: Europe and the American Moral Imagination*. Urbana: University of Illinois Press, 1961. A major work that delves deeply into the origin of the edenic theme and its many outcroppings in American thought and action.

KIRKPATRICK SALE, *Dwellers in the Land: The Bioregional Vision*. San Francisco: Sierra Club Books, 1985. Gushingly missionary in tone, but nonetheless valuable as a manifesto on behalf of a grassroots movement that seems to be gathering strength.

HENRY NASH SMITH, *Virgin Land: The American West as Symbol and Myth*. Cambridge, Mass.: Harvard University Press, 1950. The seminal work on the topic.

BRET WALLACH, *At Odds with Progress: Americans and Conservation*. Tucson: University of Arizona Press, 1991. In this finely written series of vignettes, treating nine case studies from Maine's Aroostook Valley to Kern County, California, the author gives us an unorthodox view of conservation, one that sees it as a distinctively American expression of an almost universal uneasiness about the character of the modern world.

9. Ethnic and Racial Groups

LOUIS ADAMIC, *A Nation of Nations.* New York: Harper & Row, Publishers, 1945.

JAMES P. ALLEN and EUGENE J. TURNER, *We the People: An Atlas of America's Ethnic Diversity.* New York: Macmillan, 1988. A superlative achievement in terms of both cartography and scholarship. Scores of full-color county- and state-level maps depicting every imaginable ethnic group. Although most of the graphics are based on 1980 Census data, earlier distribution and migration patterns also appear, and the text is loaded with historical information.

MOLEFI ASANTE and MARK MATTSON, *The Historical and Cultural Atlas of African Americans.* New York: Macmillan, 1991. For the first time every mappable aspect of the Afro-American experience presented, via maps and text, within a single volume.

ELAINE M. BJORKLAND, "Ideology and Culture as Exemplified in Southwest Michigan," *Annals of the Association of American Geographers* 54 (1964), 227–41. A detailed analysis of a Dutch-American region.

J. NEALE CARMAN, *Foreign Language Units of Kansas. I: Historical Atlas and Statistics.* Lawrence: University of Kansas Press, 1962. A historical-cartographic survey, county by county—nay, township by township—of the foreign-language groups of the state in staggering detail. Large mass of statistics, notes, and graphs included. The basic theme is cultural assimilation. A formidable example unlikely to be emulated by scholars in many other states.

WILLIAM M. DENEVAN, ed., *The Native Population of the Americas in 1492.* Madison: University of Wisconsin Press, 1976. The most plausible reconstructions of numbers and distributions at contact time appear in this balanced treatment of a vexed and controversial topic.

LEONARD DINNERSTEIN and DAVID M. REINERS, *Ethnic Americans: A History of Immigration and Assimilation.* New York: Dodd, Mead, 1975.

HAROLD E. DRIVER, *Indians of North America.* Chicago: University of Chicago Press, 1961. A solid, well-informed presentation that covers a score or more of major topics.

MILTON M. GORDON, *Assimilation in American Life: The Role of Race, Religion, and National Origin.* New York: Oxford University Press, 1964. An extremely useful, widely cited essay that attempts a comprehensive theoretical statement on the changing nature of social groups in the United States. Basic reading.

ANDREW M. GREELEY and WILLIAM C. MCCREADY, *Ethnicity in the United States: A Preliminary Reconnaissance.* New York: John Wiley & Sons, 1974. One of the better introductions to the subject.

A. IRVING HALLOWELL, "The Backlash of the Frontier: The Impact of the Indian on American Culture," in *Smithsonian Report for 1958.* Washington, D.C.: Smithsonian Institution, 1959. Probably the best concise statement on the subject.

OSCAR HANDLIN, *The Uprooted: The Epic Story of the Great Migrations That Made the American People.* New York: Grosset & Dunlap, 1951.

MELVILLE HERSKOVITS, *The Myth of the Negro Past.* New York: Harper & Row, Publishers, 1941. The first serious examination of the nature and extent of African cultural transfers to the United States. An epochal piece of scholarship whose implications are still not fully realized.

TERRY G. JORDAN, "Population Origin Groups in Rural Texas" (Map Supplement No. 13, scale 1:1,500,000), *Annals of the Association of American Geographers* 60:2 (1970). A splendid research effort. Now all we need are 49 maps of similar quality for the remaining states.

ALFRED L. KROEBER, *Cultural and Natural Areas of Native North America* (Publications in American Archaeology and Ethnology, Vol. 38). Berkeley: University of California Press, 1939. A masterly synthesis of what was then known or surmised about the location, population, ecology, and cultural identity of the aboriginal groups inhabiting all of North America during immediate pre-contact time.

STANLEY LIEBERSON and MARY C. WATERS, *From Many Strands: Ethnic and Racial Groups in Contemporary America.* New York: Russell Sage Foundation, 1988. A detailed demographic and sociological analysis based largely on 1980 Census data. Many original findings.

WILLIAM C. SHERMAN, *Prairie Mosaic: An Ethnic Atlas of Rural North Dakota*. Fargo: North Dakota Institute for Regional Studies, 1983. Not as detailed or ambitious as the Carman monograph on Kansas, which may well have been its inspiration, this study nonetheless should satisfy any reasonable reader's curiosity about the subject. The term "atlas" may be a misnomer for a volume with only a handful of black-and-white maps.

WILLIAM C. STURTEVANT, ed. *Handbook of North American Indians* (20 volumes projected; Volumes 4 to 11, 15 published to date). Washington, D.C.: Smithsonian Institution, 1978–. If ever a scholarly project is to be regarded as definitive, this is one of the likelier possibilities. These handsomely produced volumes written by dedicated specialists cover the subject encyclopedically, using both the topical and "tribal" approach.

GERALD D. SUTTLES, *The Social Order of the Slum: Ethnicity and Territory in the Inner City*. Chicago: University of Chicago Press, 1968. An important analysis at the micro scale of the social geography and sociology of a few square blocks of Chicago's Near West Side by a participant observer.

STEPHAN THERNSTROM, ed., *Harvard Encyclopedia of American Ethnic Groups*. Cambridge, Mass.: Belknap Press, 1980. A basic reference. Detailed essays by knowledgeable scholars.

10. The Built Landscape

ERIC ARTHUR and DUDLEY WITNEY, *The Barn: A Vanishing Landmark in North America*. Greenwich, Conn.: New York Graphic Society, 1972. A simply splendid and expert pictorial survey of the grandest of our vernacular artifacts.

GRADY CLAY, *Close-up: How to Read the American City*. New York: Praeger, 1973; reprinted, Chicago: University of Chicago Press, 1980. "A highly original, literally eye-opening 'Baedeker to the commonplace,' useful effort by an 'urban journalist and professional observer' to come to grips visually with the recent structure and changes of American cities. The chapter titles include: fixes, epitome districts, fronts, strips, beats, stacks, sinks, and turf. Imaginatively illustrated." (GBAL)

MICHAEL P. CONZEN, ed., *The Making of the American Landscape*. Boston: Unwin Hyman 1990. A set of 18 original essays focused, variably, on regions, periods, and themes. A rewarding initial conspectus of the creation of American-built landscapes past and present.

JAMES S. DUNCAN, Jr., "Landscape Taste as a Symbol of Group Identity: A Westchester County Village," *Geographical Review* 63 (1973), 334–55. A gem of an article, one that documents how social identity is revealed by cues in domestic and landscape architecture in a suburban setting.

JOEL GARREAU. *Edge City: Life on the New Frontier*. New York: Doubleday & Co., Inc., 1991. An immensely readable, fresh look by an unusually alert journalist at what has been happening to the American metropolis recently. In describing these new, unprecedented, virtually autonomous urbs mushrooming along the metropolitan rim, Garreau forces us to look deep into the puzzlements of American values and yearnings.

HENRY GLASSIE, *Folk Housing in Middle Virginia: A Structural Analysis of Historic Artifacts*. Knoxville: University of Tennessee Press, 1975. A masterful achievement. This literal and theoretical example of structuralism is an intense analysis of a handful of surviving vernacular buildings in a rural locality. Glassie enables us to peer deeply into a subconscious value system and its dynamics during a transitional period in early American history.

ALAN GOWANS, *Images of American Living: Four Centuries of Architecture and Furniture as Cultural Expression*. Philadelphia: J.B. Lippincott, 1964. A splendid, profound work—the nearest thing yet to a definitive treatment of the meaning of American architecture (and furniture).

J. FRASER HART, "Field Patterns in Indiana," *Geographical Review* 58 (1968), 450–71. Apparently the only treatment of the topic in the American literature.

J. B. JACKSON, *The Necessity for Ruins and Other Topics.* Amherst: University of Massachusetts Press, 1980. "These nine superbly written provocative essays by the founder of modern landscape analysis and criticism range widely and deeply, including such topics as European gardens and theater scenery, but the dominant concern is the American landscape. Indispensable for anyone curious about the evolution and meaning of the man-made scene." (GBAL)

————, *Discovering the Vernacular Landscape.* New Haven, Conn.: Yale University Press, 1984. "Can there ever be too much of the thoughts and prose of J. B. Jackson? Here we have fourteen of the most recent of his reflections on the evolution and implications of ordinary American landscapes. A dominant theme is the tension between the 'Landscape One' of land-based locals and the 'Landscape Two' imposed by central authority. Wisdom is not too strong a descriptor." (GBAL)

TERRY G. JORDAN, *Texas Graveyards: A Cultural Legacy.* Austin: University of Texas Press, 1982. Based upon much fieldwork in rural Texas, this may well be the best and most penetrating treatment to date of the cultural and social geography of American cemeteries. The emphasis is on ethnic differences in these older burial places, as well as their physical attributes, and the implications thereof.

FRED KNIFFEN, "Louisiana House Types," *Annals of the Association of American Geographers* 26 (1936), 179–93. The first substantial article by an American geographer on house morphology—and a durable classic.

————, "The American Covered Bridge," *Geographical Review* 41 (1951), 114–23. One of the earliest accounts of the spatial diffusion of an American innovation.

————, "Folk Housing: Key to Diffusion," *Annals of the Association of American Geographers* 55 (1965), 549–77. The culture areas of the early nineteenth century as discerned through field investigation of rural houses.

PEIRCE F. LEWIS, "The Geography of Old Houses," *Earth and Mineral Sciences* (The Pennsylvania State University) 39:5 (1970), 33–37. The only thing wrong with this statement on the historical-geographical implications of early houses is that it is too brief.

————, "The Galactic Metropolis," in Rutherford Platt and George Macinko, eds., *Beyond the Urban Fringe: Land Use Issues of Nonmetropolitan America* (pp. 23–49). Minneapolis: University of Minnesota Press, 1983. An eloquent statement on the profound restructuring of the American settlement fabric in the late twentieth century as dispersed urbanoid tissue invades and transforms so much of the countryside.

CHESTER H. LIEBS, *Main Street to Miracle Mile: American Roadside Architecture.* Boston: Little, Brown, 1985. As fine an account of the evolving morphology of commercial America as one could desire.

DAVID LOWENTHAL, "The Bicentennial Landscape: A Mirror Held Up to the Past," *Geographical Review* 67 (1977), 253–67. What do the outward signs of celebration in 1976 tell us about our changing selves, especially when we look back at 1876? Quite a bit, as Lowenthal demonstrates.

VIRGINIA and LEE MCALESTER, *A Field Guide to American Houses.* New York: Knopf, 1984. "Arguably the best in its genre, a guide that identifies and places in their historical, regional, and architectural contexts the houses built for American families (rich, poor, and in-between) in city and countryside, from the 17th Century to the present. Several maps appear amidst the profusion of illustrations." (GBAL)

D. W. MEINIG, ed., *The Interpretation of Ordinary Landscapes: Geographical Essays.* New York: Oxford University Press, 1979. "An excellent set of essays by leading scholars which examines landscapes as symbols, as expressions of cultural values, social behavior, and individual actions worked upon particular localities over a span of time." (GBAL)

LEWIS MUMFORD, *Sticks and Stones: A Study of American Architecture and Civilization.* New York: Boni & Liveright, 1924. An early and notable attempt to evaluate American architecture in terms of our civilization.

ALLEN G. NOBLE, *Wood, Brick and Stone: The North American Settlement Landscape*, 2 vols. Amherst: University of Massachusetts Press, 1984. The fullest inventory and commentary to date by a cultural geographer on folk and vernacular dwellings, barns, and other structures, in town and countryside.

WILLIAM D. PATTISON, *Beginnings of the American Rectangular Land Survey, 1784– 1800* (Research Paper No. 50). Chicago: University of Chicago, Department of Geography, 1957. A definitive essay on the first formulation of, and experiments in Ohio with, the system of land survey soon to be adopted for the entire national domain.

EDWARD T. PRICE, "The Central Courthouse Square in the American County Seat," *Geographical Review* 58 (1968), 29–60. The historical geography and diffusion of several distinct types of courthouse square.

AMOS RAPOPORT, *House Form and Culture*. Englewood Cliffs, N.J.: Prentice-Hall, 1969. A stimulating general synthesis dealing with the interaction of social, cultural, and environmental forces in building morphology. References to the American scene are only incidental.

JOHN W. REPS, *The Making of Urban America: A History of City Planning in the United States*. Princeton, N.J.: Princeton University Press, 1965. "Traces the traditions that influenced American town planning from its European background to the twentieth century. Excellent overview of American urban planning. Includes more than 300 reproductions and town plans." (GBAL)

CAROLE RIFKIND, *Main Street: The Face of Urban America*. New York: Harper & Row, Publishers, 1977. An admirable, lavishly illustrated account of the what, when, why, and how of the physical substance of the central portions of American towns and cities.

ROBERT RILEY, "Speculations on the New American Landscape," *Landscape* 24:3 (1980), 1–9. Some sharp, perceptive observations on a swiftly changing scene.

THOMAS J. SCHLERETH, ed., *Material Culture: A Research Guide*. Lawrence: University Press of Kansas, 1985. A set of eight wonderful bibliographic essays, all dealing with matters American.

———, *Cultural History and Material Culture: Everyday Life, Landscapes, Museums*. Ann Arbor, Mich.: U.M.I. Research Press, 1990. Selected essays by a particularly productive and observant student of the tangible world Americans have created. Delectable.

ERIC SLOANE, *American Barns and Covered Bridges*. New York: Wilfred Funk, 1954. Excellent line drawings, along with comments and reflections by an artist specializing in early Americana. Heartwarming.

JOHN R. STILGOE, *Common Landscape of America, 1580–1845*. New Haven, Conn.: Yale University Press, 1982. An ambitious, partially successful attempt to describe and explain the full range of the built landscape in its historical and regional dimensions.

———, *Metropolitan Corridor: Railroads and the American Scene*. New Haven, Conn.: Yale University Press, 1983. This labor of love is in a class by itself as it delineates the physical world of the railroad during its heyday along with the social and economic impact of its dominance upon the hinterland.

HARRY SWAIN and COTTON MATHER, *St. Croix Border Country*, Prescott, Wisc.: Trimbelle Press & Pierce County Geographical Society, 1968. A witty dissection of the exurban landscape. Unlikely as its parochial title and auspices may make it seem, this slim guidebook is an item of wide general interest for anyone concerned with the peripheries of American metropolitan regions.

GLENN T. TREWARTHA, "Some Regional Characteristics of American Farmsteads," *Annals of the Association of American Geographers* 38 (1948), 196–225. The only substantial effort in this direction thus far by an American geographer, and an interesting experiment.

CHRISTOPHER TUNNARD and HENRY HOPE REED, *American Skyline: The Growth and Form of Our Cities and Towns*. Boston: Houghton Mifflin Company, 1956. "The only well-rounded discussion of the evolution of the American urban landscape yet published—and a good one. Many valuable sidelights on the general historical geography of the nation." (GBAL)

WILBUR ZELINSKY, "A Toponymic Approach to the Geography of American Cemeteries," *Names* 38 (1990), 209–29. The strategy may be toponymic, but the objective in this initial try at examining American burial places at the national scale is to get at the spatial patterns of different morphological and sociocultural types.

11. *Language*

MEREDITH BURRILL, "Toponymic Generics," *Names* 4 (1956), 129–45, 226–40. The cultural and geographic significance of some of the more important generic terms in American place-names.

CRAIG M. CARVER, *American Regional Dialects: A Word Geography*. Ann Arbor: University of Michigan Press, 1987. Using the voluminous data from the DARE (*Dictionary of American Regional English*) project, the author has generated by far the best tableau to date of the various regional dialects and "dialect layers" in this country. Many maps. The analysis is deep and sophisticated.

FREDERIC G. CASSIDY, ed., *Dictionary of American Regional English. Volume I: Introduction and A–C.* Cambridge, Mass.: Harvard University Press, 1985. One of the crowning glories of American scholarship! More than 20 years in the making, this monumental dictionary exploits the results of intensive interviews in more than a thousand localities and an exhaustive documentary search. Beyond the cornucopia of linguistic information (etymology, regionalization, social and demographic correlates, etc.), this is a virtual encyclopedia on every aspect of American life and the humanized habitat. The many computerized cartograms are a revelation. With some luck, the remaining volumes could materialize by the end of this century.

CHARLES A. FERGUSON and SHIRLEY BRICE HEATH, eds., *Language in the USA.* Cambridge: Cambridge University Press, 1981. Twenty-three essays that deal effectively with most significant aspects of the topic.

JOSHUA A. FISHMAN, *Yiddish in America: Socio-Linguistic Description and Analysis* (Publication 36). Bloomington: Indiana University Research Center in Anthropology, Folklore, and Linguistics, 1965. An innovative, intriguing treatment of the linguistic geography and sociology of speakers of a dwindling language in both Eastern Europe and the United States.

——— et al., *Language Loyalty in the United States: The Maintenance and Perpetuation of Non-English Mother Tongues by American Ethnic and Religious Groups.* The Hague: Mouton, 1966. This massive, many-sided treatment by a number of contributors promises to be the most definitive on the subject.

HANS KURATH, *A Word Geography of the Eastern United States.* Ann Arbor: University of Michigan Press, 1949. "This first major project of the *Linguistic Atlas of the United States* analyzes through maps and text the spatial distribution of selected elements in the American vocabulary and offers a set of linguistic regions for the Atlantic seaboard." (GBAL)

——— and RAVEN I. MCDAVID, JR., *The Pronunciation of English in the Atlantic States.* Ann Arbor: University of Michigan Press, 1961. A sequel to the preceding item and, like it, rich in cartographic documentation.

CHARLTON LAIRD, *Language in America.* Cleveland: World Publishing, 1970. A large, ambitious work, strong in both fact and theory, that views the entire Western Hemisphere (before and after the European invasion) as a grand experiment in linguistic change through time and space.

HENRY L. MENCKEN, *The American Language: An Inquiry into the Development of English in the United States*, 4th ed. New York: Knopf, 1936. *Supplement I*, 1945; *Supplement II*, 1948. Everything you ever wanted to know about the language (including the spatial dimension and much more) by a writer who could never be dull even if he tried. Mencken was an untrained, but ardent, word-watcher—one of the last great amateur scholars in a world of galloping specialization and credentialism.

CARROLL E. REED, *Dialects of American English*. Cleveland: World Publishing 1967. A brief, semipopular, but highly serviceable introduction and survey of the subject.

RICHARD B. SEALOCK and PAULINE A. SEELY, *Bibliography of Place-Name Literature: United States and Canada*, 3rd ed. Chicago: American Library Association, 1982. Conscientiously comprehensive.

ELSDON C. SMITH, *American Surnames*. Philadelphia: Chilton, 1969. This is the only truly solid, comprehensive treatment of the subject, one that is somehow both encyclopedic and penetrating. A labor of love by another leisure-time scholar.

GEORGE R. STEWART, *Names on the Land: A Historical Account of Place-Naming in the United States*, rev. ed. Boston: Houghton Mifflin Company, 1958. This volume still stands in solitary splendor as the only worthwhile work on the subject. Scholarly and highly readable.

———, *American Place-Names: A Concise and Selective Dictionary for the Continental United States of America*. New York: Oxford University Press, 1970.

WILBUR ZELINSKY, "Some Problems in the Distribution of Generic Terms in the Place-Names of the Northeastern United States," *Annals of the Association of American Geographers* 45 (1955), 319–49.

———, "Cultural Variation in Personal Name Patterns in the Eastern United States," *Annals of the Association of American Geographers* 60 (1970), 743–69. Evidently the only geographic analysis of forenames thus far, and one of my favorite pieces of reading.

12. Religion

WHITNEY R. CROSS, *The Burned-Over District: The Social and Intellectual History of Enthusiastic Religion in Western New York, 1800–1850*. Ithaca, N.Y.: Cornell University Press, 1950.

EDWIN S. GAUSTAD, *Historical Atlas of Religion in America*, rev. ed. New York: Harper & Row, Publishers, 1976. "An expert analysis in text, map, and graph of the historical geography of the principal American denominations and some general aspects of American religion." (GBAL)

G. RINSCHEDE and S. M. BHARDWAJ, eds., *Pilgrimage in the United States*, (Geographia Religionum 5). Berlin: Dietrich Reimer Verlag, 1990. This collection of ten original essays on Catholic, Mormon, Sikh, Hindu, and secular, but quasi-religious, pilgrimages opens new vistas for the student of America's religious geography.

JAMES R. SHORTRIDGE, "Patterns of Religion in the United States," *Geographical Review* 66 (1976), 420–34.

———, "A New Regionalization of American Religion," *Journal for the Scientific Study of Religion* 16:2 (1977), 143–53.

DAVID E. SOPHER, *Geography of Religions*. Englewood Cliffs, N.J.: Prentice-Hall, 1967. The first English-language monographic treatment of the subject—and an excellent one, original in concept, and rich in ideas and fact. Several pages are devoted to the United States.

WILBUR ZELINSKY, "An Approach to the Religious Geography of the United States: Patterns of Church Membership in 1952, *Annals of the Association of American Geographers* 51 (1961), 139–93.

13. Political Behavior

MICHAEL BARONE and GRANT UJIFUSA, *The Almanac of American Politics 1990: The President, the Senators, the Representatives, the Governors: Their Records and Election Results, Their States and Districts*. Washington, D.C.: National Journal, 1989. This periodically updated compendium is a veritable treasure house of information. Contains much incidental lowdown on the human geography of the individual states and congressional districts.

STANLEY D. BRUNN, *Geography and Politics in America*. New York: Harper & Row, Publishers, 1974. A refreshingly original, many-sided approach to the American political scene. Strongly recommended.

DANIEL J. ELAZAR, *American Federalism: The View from the States*, 3rd ed. New York: Harper & Row, Publishers, 1984. "In addition to covering the theory and practice of federalism in state-local and federal-state relationships, this influential publication delves into the geography of political cultures by mapping and discussing three major modes of behavior: the individualistic, moralistic, and traditionalistic." (GBAL)

V. O. KEY, JR., *Southern Politics*. New York: Knopf, 1949. "Solid stuff on the not-so-solid South before *Brown* vs. *Topeka*. Loaded with maps, graphs, and a vast body of political and geographical savvy. If you want to learn about Southern politics in depth, start here." (P. F. Lewis)

KENNETH C. MARTIS, *The Historical Atlas of United States Congressional Districts*. New York: Macmillan, 1982. The first installment in a heroic—and essential—feat of documentary and cartographic drudgery. Martis has somehow managed to reconstruct every single congressional district for every one of our first 90-odd Congresses. Succeeding volumes (see next entry) take up voting patterns for and by Congress.

———, *The Historical Atlas of Political Parties in the United States Congress, 1789–1989*. New York: Macmillan, 1989.

KEVIN P. PHILLIPS, *The Emerging Republican Majority*. Garden City, N.Y.: Doubleday & Co., Inc., 1969. Although instantly obsolete, often superficial, and burdened with a misleading title, this exasperating mishmash does contain a superfluity of interesting, useful detail, in word and map, not only on political geography but also on the geography of settlement and ethnic groups.

RICHARD L. MERRITT, *Symbols of American Community, 1735–1775*. New Haven, Conn.: Yale University Press, 1966. How feelings of national unity germinated and sprouted during a critical period as revealed through the content analysis of place references in newspapers in four provincial metropolises. An important contribution to general political theory and quantitative behavioral studies.

14. Foodways

It has been only quite recently that American scholars, aside from nutritionists, have begun looking seriously at the cultural and social aspects of eating and drinking and associated matters in this country. In addition to the items cited below, important pioneering contributions are embedded in two wide-ranging works (noted elsewhere in this bibliographic appendix) by a pair of closet geographers: George Stewart and Rupert Vance.

American Heritage Cookbook and Illustrated History of American Eating and Drinking. New York: Simon & Schuster, 1964. "A trifle folksy in spots and regrettably undocumented, this is nonetheless the only comprehensive attempt I know to paint a broad picture of American culinary history and eating habits." (P. F. Lewis)

WARREN J. BELASCO, *Appetite for Change: How the Counterculture Took on the Food Industry, 1966–1988*. New York: Pantheon, 1989. This sprightly and definitive volume delivers just what the title promises: a sobering chronicle of how a well-nigh revolutionary turn toward natural and organic comestibles has been co-opted by our titans of commerce.

RICHARD O. CUMMINGS, *The American and His Food: A History of Food Habits in the United States*, rev. ed. Chicago: University of Chicago Press, 1941. A valiant attempt, but, alas, only a beginning.

MARGARET CUSSLER and MARY L. DE GIVE, *'Twixt the Cup and the Lip: Psychological and Socio-Cultural Factors Affecting Food Habits*. Boston: Twayne Publishers, 1952.

JOHN EGERTON, *Southern Food: At Home, on the Road, in History.* New York: Knopf, 1987. A joyous romp through the wonderful and occasionally strange world of Southern regional foodways, and a solidly informative one as well. The reader who can work his way through this account without salivating is more to be pitied than censured.

SAM B. HILLIARD, *Hog Meat and Hoecake: Food Supply in the Old South: 1840–1860.* Carbondale: Southern Illinois University Press, 1972.

HARVEY A. LEVENSTEIN, *Revolution at the Table: The Transformation of the American Diet.* New York: Oxford University Press, 1988. A solid treatment of the industrialization and commercialization of American food with a special focus on the period 1860–1930.

RICHARD PILLSBURY, *From Boarding House to Bistro: The American Restaurant Then and Now.* Boston: Unwin Hyman, 1990. At last a book-length effort to explore the restaurant phenomenon in this country. The results in this geographically oriented item are quite satisfying.

WAVERLY ROOT and RICHARD DE ROCHEMONT, *Eating in America: A History.* New York: William Morrow, 1976. This commodious volume, which largely supersedes the American Heritage production, is the best overall survey of the topic to date, but is still quite far from being definitive given its many gaps and casual wanderings.

JOE GRAY TAYLOR, *Eating, Drinking, and Visiting in the South: An Informal History.* Baton Rouge: Louisiana State University Press, 1982.

WILBUR ZELINSKY, "The Roving Palate: North America's Ethnic Restaurant Cuisines," *Geoforum* 16:1(1985), 51–72. This article has scarcity value: a first account of the topic, and an explicit approach to the theme of the transnationalization of contemporary culture.

15. Sport

JOHN F. ROONEY, JR., *A Geography of American Sport: From Calvin Creek to Anaheim.* Reading, Mass.: Addison-Wesley, 1974. "The first major American text on sport. Emphasis is on major university and professional American sports: baseball, football, and basketball. Includes a good analysis of regional patterns and spatial interactions of individual athletes, teams, and leagues." (GBAL)

——— and RICHARD PILLSBURY, *An Atlas of American Sport.* New York: Macmillan, 1992. Covers both current and historical aspects of organized sport in considerable detail. Although such major items as baseball, football, basketball and golf receive extended treatment, some 90-odd other sports are documented.

Sport Place: An International Journal of Sports Geography. Stillwater, Okla.: Black Oak Press, 1987–. Issued three times a year.

16. Folklore and Handicrafts

JAN HAROLD BRUNVAND, *The Study of American Folklore: An Introduction.* New York: W. W. Norton & Co., Inc., 1968. An excellent survey of the discipline, broad in coverage and overflowing with good bibliographies.

RICHARD M. DORSON, *American Folklore.* Chicago: University of Chicago Press, 1959. A fine general account, one that includes a generous section on modern folklore. In "Regional Folk Cultures" (pp. 74–134), the author deals with the Pennsylvania German, Spanish Southwest, Maine Coast, and Mormon areas.

———, ed., *Folklore and Folklife: An Introduction.* Chicago: University of Chicago Press, 1972. Eighteen substantive essays on specific topics and eight on methods, each with an excellent selected bibliography. The principal emphasis is generally on the United States.

HENRY GLASSIE, *Pattern in the Material Folk, Culture of the Eastern United States.* Philadelphia: University of Pennsylvania Press, 1969. An extraordinary, if eclectic and disorganized, account of an amazing range of phenomena, but mainly houses and other structures, based on extensive reading and field observation. The absence of an index and a table of contents is maddening. The bibliographic citations are uniquely valuable—if you can find them.

JOHN A. KOUWENHOVEN, *Made in America: The Arts in Modern Civilization.* Garden CIty, N.Y.: Doubleday & Co., Inc., 1948. A thoughtful essay on the chronic tension in American thought and artifice between the honest expedience of locally developed vernacular crafts and imported European traditions and standards.

E. JOAN WILSON MILLER, "The Ozark Culture Region as Revealed by Traditional Materials," *Annals of the Association of American Geographers* 58 (1968), 51–77. Verbal folklore as grist for the geographic mill.

17. Regional Studies

a. GENERAL WORKS

CONRAD M. ARENSBERG, "American Communities," *American Anthropologist* 57 (1955), 1143–60. A significant treatment of the major sociocultural regions of the country.

OTIS W. COAN and RICHARD G. LILLARD, *America in Fiction: An Annotated List of Novels That Interpret Aspects of Life in the United States, Canada, and Mexico,* 5th ed. Palo Alto, Calif.: Pacific Press, 1967.

JOEL GARREAU, *The Nine Nations of North America.* Boston: Houghton Mifflin Company, 1981. "A successful, highly readable and semi-popular account by a journalist of 'the way North America really works'—at least in regional terms. Garreau maps, describes, and interprets nine semi-autonomous 'nations' within the United States, Canada, Mexico, and the Caribbean that cohere socially, culturally, and economically with little regard for state, provincial, or other political boundaries: in essence functional-vernacular geography." (GBAL)

RAYMOND D. GASTIL, *Cultural Regions of the United States.* Seattle: University of Washington Press, 1975. The competition. Out of print, of course.

A. R. MANGUS, *Rural Regions of the United States,* Washington, D.C.: Works Progress Administration, 1940. Figure 1 (facing, p. 4) is a map entitled "Rural-Farm Cultural Regions and Sub-Regions," one that depicts 32 regions and 218 subregions based upon population characteristics.

ANN R. MARKUSEN, *Regions: The Economics and Politics of Territory.* Totawa, N.J.: Rowman & Littlefield, 1987. An enterprising economist's effort to come to grips with the stubborn reality of regional differences and their economic and political implications, but recognition of the cultural factor is still far from adequate.

DONALD W. MEINIG, "The American Colonial Era: A Geographic Commentary," *Proceedings of the Royal Geographical Society, South Australia Branch* (1957–58), 1–22. The origins, nature, and significance of the three principal colonial culture hearths stated pithily and effectively.

ROBERT D. MITCHELL, "The Formation of Early American Cultural Regions: An Interpretation," in James R. Gibson, ed., *European Settlement and Development in North America* (pp. 66–90). Toronto: University of Toronto Press, 1978. As sound and detailed a presentation as has appeared thus far of the evolution of culture areas in the eastern third of the country.

MICHAEL STEINER and CLARENCE MONDALE, *Region and Regionalism in the United States: A Source Book for the Humanities and Social Sciences.* New York: Garland, 1988. All 14 sections of this remarkably far-ranging, liberally annotated, topically arranged bibliography merit notice by the student of American cultural geography, but the richest pickings are in the many pages devoted to geography and history.

WILBUR ZELINSKY, "Selfward Bound? Personal Preference Patterns and the Changing Map of American Society," *Economic Geography* 50 (1974) 144–79. The sociocultural and psychological regionalization of the country as revealed by analysis of state-level data on membership in voluntary associations and readership of special-interest magazines.

————, "North America's Vernacular Regions," *Annals of the Association of American Geographers* 70 (1980), 1–16. The regionalization of the United States and Canada from a nontraditional perspective: what labels the residents of metropolitan areas apply to their corner of the world as indicated by names of enterprises listed in telephone directories.

CARLE C. ZIMMERMAN and R. E. DU WORS, *Graphic Regional Sociology (A Study in American Social Organization)*. Cambridge, Mass.: Phillips Book Store, 1952. A curious mixture of syllabus and essay, cliché and fresh insight. The sociology, economy, and culture of seven regions are presented: the South; Appalachian-Ozark Region; Northeastern-Urban Industrial Region; Corn Belt; Wheat Belt; Arid West; and Pacific Region.

b. NEW ENGLAND

HANS KURATH, *Handbook to the Linguistic Atlas of New England*. Providence, R.I.: Brown University Press, 1943. Ventures well beyond the realm of language to offer a sound historical, geographic, social, and cultural introduction to the region.

SUMNER CHILTON POWELL, *Puritan Village: The Formation of a New England Town*. Middletown, Conn.: Wesleyan University Press, 1963. An exemplary analysis of how the social and physical attributes of the town were crystallized from elements imported from Old England.

LOIS (KIMBALL) MATHEWS ROSENBERRY, *The Expansion of New England: The Spread of New England Settlement and Institutions to the Mississippi River, 1620–1865*. Boston: Houghton Mifflin Company, 1909.

WALTER M. WHITEHILL, *Boston: A Topographical History*, 2nd ed. Cambridge, Mass.: Harvard University Press, 1968. "The classic description of the present topography of the inner part of Boston city and the processes by which today's geographical landscape was shaped. Profusely illustrated with photographs and maps." (GBAL)

RICHARD W. WILKIE and JACK TAGER, eds., *Historical Atlas of Massachusetts*. Amherst: University of Massachusetts Press, 1991. This splendid production, one that sets lofty new standards for state and regional historical atlases, deals with a wide range of topics within a state as pivotal as any in the shaping of American society and culture. Detailed coverage of politics, ethnicity and race, the built environment, communications, medical geography, and "women and society," inter alia.

JOSEPH S. WOOD, "Village and Community in Early Colonial New England," *Journal of Historical Geography* 8 (1982), 333–46. After meticulous scrutiny of all available maps and other documentation, the author destroys the myth that colonial New England settlement was overwhelmingly concentrated in clustered villages—a phenomenon that materialized only toward the end of the eighteenth century.

c. THE MIDLAND

E. DIGBY BALTZELL, *Puritan Boston and Quaker Philadelphia: Two Protestant Ethics and the Spirit of Class Authority and Leadership*. New York: Free Press, 1979. As the title suggests, this clever historian has perpetrated some admirable psycho-history by showing how the quite different collective mind-sets of the elites of these two highly influential cities were molded by their theological heritage. The social, economic, and political consequences have been noteworthy. The documentation is fascinating.

DAVID CUFF, EDWARD K. MULLER, WILLIAM J. YOUNG, WILBUR ZELINSKY, and RONALD F. ABLER, eds., *The Atlas of Pennsylvania*. Philadelphia: Temple University Press, 1989. A model in its genre. Social, cultural, and historical items receive more than their customary share of the limelight.

STEVENSON W. FLETCHER, *Pennsylvania Agriculture and Country Life, 1640–1840*. Harrisburg: Pennsylvania Historical and Museum Commission, 1950. An amazingly detailed and comprehensive book.

JOSEPH W. GLASS, *The Pennsylvania Culture Region: A View from the Barn*. Ann Arbor, Mich.: U.M.I. Research Press, 1986. The region delineated, described, and discussed on the basis of farmstead and barn characteristics carefully noted in the field.

JEAN GOTTMANN, *Megalopolis: The Urbanized Northeastern Seaboard of the United States.* New York: Twentieth Century Fund, 1961. "The classic overview of the urban region that stretches from Washington, D.C., to Boston. Supporting data are now old, but the general material is surprisingly current. Physical, historical, social, political and economic aspects are discussed individually in detail and then reviewed in their regional or megalopolitan context." (GBAL)

JOHN A. HOSTETLER, *Amish Society.* Baltimore: The Johns Hopkins University Press, 1963. Perhaps no other population of comparable minuteness has stimulated so much general interest or so many publications, good and bad. Hostetler, who writes with the special vision of an insider, provides perhaps the most satisfying overall survey of this peculiar community.

JAMES T. LEMON, *The Best Poor Man's Country: A Geographical Study of Early Southeastern Pennsylvania.* Baltimore: The Johns Hopkins University Press, 1972. "Prize-winning study of the settlement and transformation of southeastern Pennsylvania between 1680 and 1800. Examines the physical background, characteristics of early settlers, population patterns, land tenure, rural and urban settlement, and regional variations. Argues that a liberal individualistic philosophy was the key to the region's development." (GBAL)

SHERRY H. OLSON, *Baltimore: The Building of an American City.* Baltimore: The Johns Hopkins University Press, 1980. "Examines the physical growth of Baltimore from the early eighteenth century until the present. Growth and expansion are associated with long-term business and building cycles and their effects on morphology and internal structure of the city. Profusely illustrated." (GBAL)

JOHN H. THOMPSON, ed., *The Geography of New York State.* Syracuse, N.Y.: Syracuse University Press, 1966. This collaborative work sets a standard not yet rivaled by any other publication on the geography of an American state. Of special interest are the three chapters on historical geography by Donald W. Meinig.

THOMAS J. WERTENBAKER, *The Founding of American Civilization: The Middle Colonies.* New York: Charles Scribner's Sons, 1938.

WILBUR ZELINSKY, "The Pennsylvania Town: An Overdue Geographical Account," *Geographical Review* 67 (1977), 127–47. A distinctive settlement type described, analyzed, celebrated, and partially explained. Based primarily on field observation.

d. THE SOUTH

HARRIET L. ARNOW, *Hunter's Horn.* New York: Macmillan, 1949. Good regional fiction is a must for the serious student of cultural geography; and here is a logical choice if only a single sample must be cited. This story of a remote community in Kentucky's Cumberland Valley ca. 1941 contains some superlative ethnography, history, regional sociology, physical and social geography, all delivered in first-rate prose and a hypnotically gripping narrative.

JEAN GOTTMANN, *Virginia in Our Century.* Charlottesville University of Virginia Press, 1969. A multifaceted regional study by the noted French geographer. This is virtually unequaled as a full-dimensional geographical monograph on an American region.

J. FRASER HART, *The Southeastern United States.* New York: Van Nostrand Reinhold, 1967. A capsule regional geography, and one devoting a relatively large share of its attention to matters social and cultural.

TERRY G. JORDAN, *German Seed in Texas Soil: Immigrant Farms in Nineteenth Century Texas.* Austin: University of Texas Press, 1966.

———, "The Imprint of the Upper and Lower South on Mid-Nineteenth-Century Texas," *Annals of the Association of American Geographers* 57 (1967), 667–90. The demographic and cultural contributions of the two principal streams of migration into Texas.

————, "The Texas Appalachia," *Annals of the Association of American Geographers* 50 (1970), 409–27. How a replica of Appalachia and Ozarkia was created in Texas by bearers of their culture.

PEIRCE F. LEWIS, *New Orleans: The Making of an Urban Landscape*. Cambridge, Mass: Ballinger, 1976. A sterling example of how to do a well-rounded geographical treatment of a fascinating city, or any city for that matter. The writing is exceptional.

DONALD W. MEINIG, *Imperial Texas: An Interpretive Essay in Cultural Geography*. Austin: University of Texas Press, 1968. Another Meinig classic. A concise, penetrating, highly original study of how the populations and culture areas of the state have taken shape over time and space.

HARRY ROY MERRENS, *Colonial North Carolina in the Eighteenth Century: A Study in Historical Geography*. Chapel Hill: University of North Carolina Press, 1964. "A study of the changing geography of North Carolina before the American Revolution, especially between 1750 and 1775. Examines administrative and political organization, environmental perception, population, economic activities, urban development, and decentralized trade." (GBAL)

ROBERT D. MITCHELL, *Commercialization and Frontier: Perspectives on the Early Shenandoah Valley*. Charlottesville: University Press of Virginia, 1977. "Examines the Shenandoah Valley of Virginia during the eighteenth century from the perspective of the relationships between economic development, social stratification, and settlement structure. Emphasizes environmental factors, population and settlement, land acquisition and speculation, social structure, pioneer economy, commercial agriculture, and related transport, urban, and trading networks." (GBAL)

FREDERICK LAW OLMSTED, *The Cotton Kingdom: A Traveller's Observations on Cotton and Slavery in the American Slave States*, ed. by Arthur M. Schlesinger. New York: Knopf, 1953; originally published 1861. Firsthand description and interpretation of the highest order. The work of this gifted social critic and landscape architect remains a basic source for the historical geography of the South.

KARL B. RAITZ and RICHARD ULACK, *Appalachia: A Regional Geography, Land, People, and Development*. Boulder, Colo.: Westview Press, 1984.

JOHN SHELTON REED, *The Enduring South; Subcultural Persistence in Mass Society*. Lexington, Mass.: Lexington Books, 1972. Sociologist Reed knows whereof he speaks when he discourses about his native South and argues persuasively that, contrary to so many expectations, this changing region shows little inclination to lose its special identity and still remains a sort of nation apart. He pursues this theme in other venues. Even if he tried, Reed could never be dull or boring.

————, *One South: An Ethnic Approach to Regional Culture*. Baton Rouge: Louisiana State University Press, 1982. See preceding entry.

CÉCYCLE TRÉPANIER, *French Louisiana at the Threshold of the 21st Century* (Projet Louisiane, Monographie, No. 3). Quebec: Université Laval, Département de Géographie, 1989. An intensive look at the cultural geography of this quite special region and its dynamics. The author is not sanguine about the prospects for its continued vitality.

RUPERT B. VANCE, *Human Geography of the South: A Study in Regional Resources and Human Adequacy*. Chapel Hill: University of North Carolina Press, 1935. After more than five decades, this treatment of the region by a sociologist has not been superseded. Contains material on social and cultural items, including diet, not found in the ordinary geography or sociology text.

CHARLES R. WILSON and WILLIAM FERRIS, eds., *The Encyclopedia of Southern Culture*. Chapel Hill: University of North Carolina Press, 1989. A truly stunning achievement! The 1700-odd two-column pages of this mighty tome (of much more than parochial interest) cover every imaginable aspect of the region's life and culture broadly defined and many topics that would never occur to even the most imaginative reader. And then there is that optimum number of great illustrations. This is the only encyclopedia I have ever encountered that is a genuine page-turner. The fact that no other American culture area has, or is even contemplating, any such publication testifies eloquently to Dixie's robustness.

e. THE MIDDLE WEST

JOHN R. BORCHERT, *America's Northern Heartland: An Economic and Historical Geography of the Upper Midwest*. Minneapolis: University of Minnesota Press, 1987. Masterful! This exemplary example of regional exposition and interpretation is the culmination of many years of study and active involvement.

JOSEPH W. BROWNELL, "The Cultural Midwest," *Journal of Geography* 39 (1960), 81–85. Apparently the first attempt to delineate the outer limits of the region.

DAVID BUISSERET, *Historic Illinois from the Air*. Chicago: University of Chicago Press, 1990. A smashing combination of early cartography and iconography, informative text, recent low oblique aerial photography, and other goodies. "The best historic landscape interpretation yet!" (D. W. Meinig)

JOHN C. HUDSON, *Plains Country Towns*. Minneapolis: University of Minnesota Press, 1985. A detailed examination of the origin, dynamics, configuration, and other facets of the human geography of the small, largely railroad-spawned towns of the Upper Midwest, mainly in the Dakotas and Minnesota.

———, "North American Origins of Middlewestern Frontier Populations," *Annals of the Association of American Geographers* 78 (1988), 395–413. A reconstruction, via early Census materials, of the provenience of the first waves of settlers in the region. In this important article, Hudson clearly documents four parallel, partially overlapping streams of migrants from four Atlantic Coast culture hearths (including the St. Lawrence Valley).

DAVID GRAHAM HUTTON, *Midwest at Noon*. Chicago: University of Chicago Press, 1946. ". . . combines historical perspectives, economic analyses, personal observations, psychological characterizations, and social interpretation in about equal measure to make a rich and sympathetic volume which resists summary." (U.S. Library of Congress, *A Guide to the Study of the United States of America*)

FRANK R. KRAMER, *Voices in the Valley: Mythmaking and Folk Belief in the Shaping of the Middle West*. Madison: University of Wisconsin Press, 1964. The role of myth in the Middle West (and Huronia) and the parental New England, Pennsylvanian, and southern regions. Pretentious, overwritten, and often incoherent, but with occasional flashes of brilliance. The myth of the homestead is one of the central items.

HAROLD M. MAYER and RICHARD C. WADE, *Chicago: Growth of a Metropolis*. Chicago: University of Chicago Press, 1969. A unique, masterly visual-verbal-cartographic study of growth, in form and process, and of a multitude of urban phenomena, including vernacular architecture and ethnic geography, along with the usual economic topics.

JAMES R. SHORTRIDGE, *The Middle West: Its Meaning in American Culture*. Lawrence: University Press of Kansas, 1989. A native son explores the origins of the idea of a Middle West, the shifting meaning, perceptions, and physical locus thereof, and attitudes toward the region by outsiders and others. Although not the definitive account of the Middle West in all its geographic actuality, it is a fine study in its own terms.

f. THE WEST

DANIEL D. ARREOLA, "Mexican American Housescapes," *Geographical Review* 78 (1988), 299–315.

REYNER BANHAM, *Los Angeles: The Architecture of Four Ecologies*, New York: Harper & Row, Publishers, 1971. Amidst a large, growing literature on this most fascinating and perplexing of American cities, this work remains outstanding for the depth and intelligence of its observations. The central concern may be architectural, but the discussion reaches far into other areas.

GIOVANNI BRINO, "Verso la Metropoli Post-Urbana: Los Angeles," *Comunita* No. 172 (1974), 316–414. Los Angeles as perceived by a sensitive, articulate Italian observer in a longish essay, effectively combining word and illustration. One of the earlier items to recognize and luxuriate in the postmodern strangeness of the place.

MICHAEL W. DONLEY, STUART ALLAN, PATRICIA CARO, and CLYDE P. PATTON, *Atlas of California*. Culver City, Calif.: Pacific Book Center, 1979. A colorful, sophisticated atlas loaded with all manner of worthwhile data, and one that ranks within the top ranks of state atlases.

RICHARD V. FRANCAVIGLIA, *The Mormon Landscape: Existence, Creation, and Perception of a Unique Image in the American West*. New York: AMS Press, 1978. "Examines the significance of the Mormons in creating a distinctive landscape in the American west, associated especially with rural-village settlement and irrigated agriculture." (GBAL)

JOHN L. LANDGRAF, *Land-Use in the Ramah Area of New Mexico: An Anthropological Approach to Areal Study* (Papers of the Peabody Museum of American Archaeology and Ethnology, Vol. 42, No. 1), Cambridge, Mass.: Harvard University Press, 1954. An excellent study of the area, the various peoples (Anglos, Texans, Mormons, Hispanic-Americans, and Navaho), their relationships to the land and to one another.

G. MALCOLM LEWIS, "Regional Ideas and Reality in the Cis-Rocky Mountain West," *Transactions and Papers, Institute of British Geographers*, Publication 38 (1966), 135–50. The Great Plains as a mental concept and as a mappable reality.

WILLIAM G. LOY, *Atlas of Oregon*. Eugene: University of Oregon, 1976. A lovely combination of geographic scholarship and fine cartographic craft. One of the best state atlases.

DONALD W. MEINIG, "The Mormon Culture Region: Strategies and Patterns in the Geography of the American West, 1847–1964," *Annals of the Association of American Geographers* 55 (1965), 191–220. In addition to being a masterly synthesis of the historical geography of Mormon population and culture, this article also proposes an interesting, influential general model for culture areas.

———, *The Great Columbia Plain: A Historical Geography, 1805–1910*. Seattle: University of Washington Press, 1968. "A detailed chronological reconstruction of eastern Washington and adjacent Oregon during the nineteenth century." (GBAL)

———, *Southwest: Three Peoples in Geographical Change, 1600–1970*. New York: Oxford University Press, 1971. A study of New Mexico and Arizona. Although it stands brilliantly on its own merits, it is most usefully read in conjunction with Meinig's earlier volume on Texas, which it resembles in both format and quality.

RICHARD L. NOSTRAND, "The Hispanic-American Borderland: Delimitation of an American Culture Region," *Annals of the Association of American Geographers* 60 (1970), 638–61.

JOHN W. REPS, *Cities of the American West: A History of Frontier Urban Planning*. Princeton, N.J.: Princeton University Press, 1979.

E. W. SOJA, "Taking Los Angeles Apart: Some Fragments of a Critical Human Geography," *Environment and Planning D; Society and Space*, 4 (1986), 255–72. A neo-Marxist, postmodern look at one of the most postmodern of metropolises. A richly informative overview and dissection of a huge, complex, and bewildering place whose dynamics have so much to tell us about both America and the larger world.

WILLIAM L. THOMAS, JR., ed., "Man, Time, and Space in Southern California," *Annals of the Association of American Geographers* 49:3, Part 2 (1959). A symposial effort: five excellent essays, plus critical commentaries.

WALTER PRESCOTT WEBB, *The Great Plains*. Lexington: Ginn & Company, 1931. "Classic book by a geographically and ecologically attuned historian; has evoked much subsequent work and discussion by scholars, including geographers. Conceives of the Great Plains not as commonly delimited on physiographic maps, but as a more extensive environment wherein settlers from the humid East had to modify their institutions drastically in response to a combination of relatively level terrain, treelessness, and moisture deficiencies." (GBAL)

18. *Miscellaneous and Unclassifiable*

CHARLES S. AIKEN, "Faulkner's Yoknapatawpha County: Geographical Fact into Fiction," *Geographical Review* 67 (1977), 1–21 and "Faulkner's Yoknapatawpha County: A Place in the American South," *Geographical Review* 79 (1979), 331–48. This fine pair of essays nicely exemplifies a relatively new genre in geographic and literary scholarship: how familiarity with a major author's writings and its rootedness in a specific locale can enrich our understanding of both.

DAVID T. BAZELON, *Power in America: The Politics of the New Class.* New York: New American Library, 1967. As one reviewer noted, this is indeed "a book of subtle contradictions . . . brilliant, provocative, and self-consciously seductive." It charges off inconclusively in many interesting directions, but the underlying theme is a powerful one: America as a shallow society of confused individualists lacking in genuine community.

WARREN JAMES BELASCO, *Americans on the Road: From Autocamp to Motel, 1910–1945.* Cambridge, Mass.: MIT Press, 1979. As satisfying and readable an account as one could desire of the evolving motoring practices of Americans during the formative period and the accompanying development of roadside features.

DANIEL J. BOORSTIN, *The Americans: The National Experience.* New York: Random House, 1965. An assemblage of essays—brilliant and original for the most part—dealing with some major aspects of American life in the early nineteenth century, among them the versatility of New Englanders, transience as a basic theme in American existence, boosterism, "the vagueness of the land" (that is, the spatial ambiguity of America), language and speech. For sheer refreshment, try the sparkling essay on the American hotel (pp. 134–47).

RICHARD D. BROWN, *Modernization: The Transformation of American Life, 1600–1865.* New York: Hill & Wang, 1976. Far from definitive, but still the best single-volume treatment to date of this mighty subject.

DEBORAH M. BUREK, ed., *Encyclopedia of Associations, 1991,* 25th ed., 4 vols, Detroit: Gale Research Co., 1990. If you need convincing that America is a nation of joiners, just dip into this fascinating reference work. It is a glorious panorama of the country in all its diversity and occasional weirdness.

GEORGE O. CARNEY, ed., *The Sounds of People and Places: Readings in the Geography of American Folk and Popular Music.* Lanham, Md.: University Press of America, 1987. The geography of American music finally emerges in book form. Contains worthwhile essays that also happen to suggest how much more remains to be done.

JAMES H. CASSEDY, *Demography in Early America: Beginnings of the Statistical Mind.* Cambridge, Mass.: Harvard University Press, 1969. Significant not only for facts and chronicling of demographic thought, knowledge, theory, and programs, but, as the subtitle indicates, as a case study in the budding of the "statistical mind." Transdemographic implications.

ERIC FISCHER, *The Passing of the European Age: A Study of the Transfer of Civilization and Its Renewal in Other Continents,* rev. ed. Cambridge, Mass.: Harvard University Press, 1948.

EDWARD T. HALL, *The Hidden Dimension.* Garden City, N.Y.: Doubleday & Co., Inc., 1966. A stimulating exploration, albeit sometimes fuzzy, of the more intimate spatial-sensory worlds of human beings. In this pioneering effort in "proxemics" (the perception and culturally conditioned use of personal space), Hall deals only briefly or tangentially with Americans, but this volume holds interest for students of any cultural group.

CHRISTOPHER HITCHENS, *Blood, Class, and Nostalgia: Anglo-American Ironies.* New York: Farrar, Straus & Giroux, 1990. American thought and behavior cannot be fully fathomed unless we factor in that "special relationship" with Great Britain. Journalist Hitchens, who has lived and worked in both countries, offers as fully rounded and witty an analysis of this love-hate bonding over the decades as we are ever likely to see. Even though the author is reasonably fair, there is much here to cause Anglophiles (and especially Churchill groupies) to bridle.

MARK HOLLOWAY, *Heavens on Earth: Utopian Communities in America, 1680–1880,* 2nd ed. New York: Dover Publications, 1966.

MICHAEL KAMMEN, *Season of Youth: The American Revolution and the Historical Imagination.* New York: Knopf, 1978. As the central, generative episode in American history, the Revolution is the most unforgettable of facts. In this rich and nourishing monograph, historian Kammen is concerned not with the event itself but with its legacy in thought and memory, and how radically transformed perception and celebration have become after two centuries.

HANS KOHN, *American Nationalism: An Interpretive Essay.* New York: Crowell-Collier-Macmillan, 1957. A broad-ranging, generally insightful essay on five facets of the origin and character of American nationalism: its roots; the unique familial link with Britain and its culture; sectionalism; the significance of large-scale immigration; and the United States as a world power. Marred somewhat by the Cold War phobias of the period.

JOHN LEIGHLY, ed., *Land and Life: A Selection from the Writings of Carl O. Sauer.* Berkeley: University of California Press, 1963. Some key essays by the most influential of all American cultural geographers.

ALLAN I. LUDWIG, *Graven Images: New England Stonecarving and Its Symbols, 1650–1815.* Middletown, Conn.: Wesleyan University Press, 1966. A splendidly produced volume that converges on its subject from several angles: the basic theory of visual symbols; Puritan theology; the burial customs of Puritans and English; and the historical-spatial-stylistic analysis of New England gravestones. A rare treatment of the cultural geography of a folk art in North America.

ROBERT O. MEAD, *Atlantic Legacy: Essays in American-European Cultural History.* New York: New York University Press, 1969. An account, roughly chronological, of the continuing transatlantic exchange of influences at the higher cultural and psychological levels.

ROBERT MEREDITH, ed., *American Studies: Essays on Theory and Method.* Columbus, Ohio: Charles E. Merrill Publishing Co., 1968. An excellent, well-edited collection of papers dealing with methodological and substantive issues in American Studies.

JOSHUA MEYROWITZ, *No Sense of Place: The Impact of Electronic Media on Social Behavior.* New York: Oxford University Press, 1985. The pervasiveness of mass electronic communication, and television in particular, has radically transformed our perceptions of space and place and most forms of social interaction. We have here a masterly account of the phenomenon, one that pervades nearly the entire world, but the United States perhaps more than any other country.

PERRY MILLER, *The Life of the Mind in America, From the Revolution to the Civil War.* New York: Harcourt Brace Jovanovich, 1965. Two-and-a-fraction parts of Miller's unfinished final opus. Highly erudite, literate prose. Book One deals with evangelism and revivalism; Book Two considers "The Legal Mentality" and the attitude of Americans toward law and lawyers.

LEWIS MUMFORD, *The Myth of the Machine: Technics and Human Development.* New York: Harcourt Brace Jovanovich, 1967 and *The Myth of the Machine: The Pentagon of Power.* New York: Harcourt Brace Jovanovich, 1970. These two volumes climax a scholarly career that in its depth and breadth defies classification or capsule description. This is a most eloquent statement concerning the interplay among myth, technology, and social behavior from prehistoric times to the present, and provides a rich backdrop for any analysis of a contemporary community.

NEIL POSTMAN, *Amusing Ourselves to Death: Public Discourse in the Age of Show Business.* New York: Viking, 1985. Postman covers much the same territory handled by Meyrowitz, while focusing sharply on the impact of TV on public discourse and, specifically, politics, both loosely and broadly defined. A sobering—nay, frightening—account.

KIRKPATRICK SALE, *Power Shift: The Rise of the Southern Rim and Its Challenge to the Eastern Establishment.* New York: Random House, 1975. A strident, almost hysterical, schizoid account that overstates the demographic-economic-political case on behalf of the "Rim" (now better known as the Sun Belt), then sounds loud alarms about the shift. Despite excessive generalization and simplistic journalism, much interesting material on regional psychology.

GEORGE R. STEWART, *American Ways of Life.* Garden City, N.Y.: Doubleday & Co., Inc., 1954. "A rare instance of an angel rushing in where fools fear to tread. Stewart writes about all sorts of peculiarly American things (including almost the only decent thing on American diet that I know about) and does it very well. The book is now out of print—a literary and scholarly catastrophe." (P. F. Lewis)

ALAN TRACHTENBERG, *Brooklyn Bridge: Fact and Symbol.* New York: Oxford University Press, 1965. An eminently satisfying exploration of the history and meaning of one of the most influential artifacts wrought upon the American continent. This particular bridge does have a great deal to tell us about ourselves, as some poets and painters have discovered.

EDWARD L. ULLMAN, "Amenities as a Factor in Regional Growth," *Geographical Review* 44 (1954), 119–32. A manifesto concerning the importance of transeconomic factors in the demographic and social development of the United States, one whose implications came only belatedly to be fully appreciated.

LAURENCE VEYSEY, *The Communal Experience: Anarchist and Mystical Countercultures in America.* New York: Harper & Row, Publishers, 1973. A worthwhile account of nineteenth and twentieth century communal experiments.

LOUIS R. WILSON, *The Geography of Reading: A Study of the Distribution and Status of Libraries in the United States.* Chicago: American Library Association and University of Chicago Press, 1938. The only study of its kind—and one that has been oddly ignored by the scholarly public. Bookstores and literacy, as well as libraries, receive much attention in this intriguing monograph.

WILLIAM WOODRUFF, *America's Impact on the World: A Study of the Role of the United States in the World Economy, 1750–1970.* New York: John Wiley & Sons, 1975. Can the impact of American culture be far behind? We won't really know until someone undertakes the task of finding out in a volume as solid as this one.

JOHN K. WRIGHT, *Human Nature in Geography: Fourteen Papers, 1925–1965.* Cambridge, Mass.: Harvard University Press, 1966. "Fourteen imaginative papers . . . touching on marvels, discoveries, philosophies, organization, and human awareness of spatial diversity." Deals with American topics, among others. A writer of rare charm, versatility, and erudition with whom every student of the American scene should be acquainted.

Index

A

Abbott, C., 179
Abler, R. F., 79, 204
Aboriginal population:
cultural impact of, 6, 14–17, 130, 196
cultural revival, 164
distribution of, 30 (*map*), 114
policy toward, 15–16, 24
size of, 15
Acadiana, 174, 180, 206
Adamic, L., 195
Adams, J., 151
Adventists, 94, 99
Afro-Americans, 18–20, 25, 162, 195
cultural revival, 164
culture, 18–20
distribution, 28, 29, 30 (*map*), 77
murals, 171
periodicals, 175
Agee, J., 88
Agriculture, 16–17
Aiken, C. S., 209
Alkjaer, J. L., 157

Allan, S., 208
Allen, J. P., 195
Americanism, 96–97
American Revolution, 13, 121, 210
American Studies, 37, 210
Amish, 11, 205 (*see also* Mennonites)
Amphibious Regions, 136–37
Animal Rights Movement, 162
Antiques, 177
Apartment buildings, 90
Appalachians, 11, 123, 127, 206
Applebaum, E. L., 176
Architecture, 83, 150, 151, 174, 177
Arensberg, C. M., 37, 120, 143, 151, 189, 203
Argentina, 10, 39
Arizona, 22, 110, 114, 130, 137
Armenians, 29
Arnow, H. L., 152, 205
Arreola, D. D., 169, 207
Arthur, E., 196
Artifactual aspects of culture, 73–74
Asante, M., 195

Asians, 23, 162, 163, 171
Assembly of God, 99
Assimilation, 32, 163
Atlanta, Ga., 138
Atlases, 144, 188–89
Augelli, J. P., 9
Austin, Tex., 174
Australia, 9, 45
Austria, 163
Axtell, J., 192

B

Bachelard, G., 88
Bailyn, B., 192
Baltimore, Md., 133, 137, 205
Baltzell, E. D., 204
Banham, R., 207
Baptists, 12, 98
Barbecue, 167
Barns, 84, 100–101, 174, 196
Barone, M., 200
Basques, 29, 164
Bathrooms, 93
Battlefield parks, 147, 153
Bazelon, D. T., 44, 153, 209
Belasco, W. B., 162, 201, 209
Belgians, 25
Berger, M., 36
Bermuda, 9
Bernal, J. D., 69
Berton, P., 148
Bhardwaj, S. M., 200
Bible Belt, 180
Bibliographies, 186–87
Big Bend, Tex., 180
Billington, R. A., 82, 144, 192
Bjorklund, E. M., 28, 99, 195
Blacks, *see* Afro-Americans
Blake, P., 102, 145, 193
Blue jeans, 158
Blumer, H., 82
Bogue, D. J., 26, 27, 56
Bohemias, Latter-Day, 138–39
Boorstin, D. J., 37, 59, 91, 121, 154, 209
Borchert, J. R., 58, 207

Born, W., 145, 193
Boston, Mass., 28, 90, 95, 121, 133, 138, 151, 166, 204
Boulder, Colo., 181
Boundaries, cultural significance of:
  internal, 113–14, 121
  international, 113, 138
Bourbon, 17, 85
Bracey, H. E., 10
Brazil, 7, 10, 19, 39
Brino, J., 208
British, 23, 24–25, 29, 32
British Columbia, 133
British cultural heritage, 4
British West Indies, 9, 25, 35, 121, 123
Brooklyn, N.Y., 11
Brooklyn Bridge, 211
Brown, R. D., 209
Brown, R. H., 82, 189, 192
Brownell, J. W., 128, 180, 207
Bruk, S. I., 28
Brunn, S. D., 148, 200
Brunvand, J. H., 107–8, 150, 202
Brush, J. E., 10
Buisseret, D., 207
Bumper stickers, 170–71
Bungalows, 174
Burek, D. M., 209
"Burned-Over District", 84, 107, 117
Burrill, M. F., 105, 109
Bush Negroes, 11
Buttons, 170

C

Cajuns, 21, 164, 175
California, 20, 21, 26, 58, 85, 137, 174, 208 (*see also* Southern California)
Campbell, R. D., 38
Canada, 6, 9, 10, 23, 25, 26, 35, 39, 45, 148, 175, 188
  culture areas of, 128
Canadians, 163

Cape Cod House, 174
Cappon, L. J., 188
Carman, J. N., 26, 134, 195
Carney, G. O., 209
Caro, P., 208
Caro, R., 165
Carter, G. F., 4, 79
Carver, C. M., 199
Cassedy, J. H., 154, 209
Cassidy, F. G., 107, 199
Castells, M., 161
Castro district, 181
Catholicism, Roman, 95, 98, 159
Catskills, 28
Cemeteries, 49, 101, 144, 160,
    197, 199
Central California Culture Area,
    130, 131–32
Charleston, S.C., 83, 133
Chesapeake Bay, 24, 83, 122
Chicago, Ill., 11, 77, 85, 129,
    133, 138, 182, 196, 207
Chicanos, *see* Spanish-Ameri-
    cans
Chinese, 29, 30 (*map*), 43–44,
    163, 191
Church buildings, 89, 95
Church denominations, 94–100,
    200
Church of the Nazarene, 99
Cincinnati, Ohio, 78, 101, 133,
    138
Cincinnatus, Order of, 147
Circulation, 56–57, 157
Cities:
    attitudes toward, 49, 167
    ruralization, 167
Civil War, 18
Clay, G., 196
Clothing, 167, 170, 177
Clough, W.D., 145, 194
Coan, O.W., 8, 203
Cold War, 147
Colonial Revival, 152
Colorado, 22, 115, 129, 130, 137
Colorado Piedmont Culture
    Area, 132–33
Communal colonies, 138–39,
    211

Communications, 8, 50, 51, 80,
    153–55, 165, 210
Conestoga Wagon, 83
Congregationalists, 12, 98
Connecticut, 47
Constitution, 160
Consumption patterns (*table*),
    106
Convergence:
    cultural, 85–88, 110, 182–83
    rural-urban, 164–66
Conzen, M. P., 196
Corey, K. E., 101
Cornish, 20, 29
Cosgrove, D. E., 169
Cotton, 17, 61
Cotton Belt, 123
County fairs, 107
Courthouses, 48, 102
Cox, E. G., 36
Cox, K. R., 107, 120
Crevecoeur, J. H., St. J. de, 3
Crime, 148
Croats, 164
Cronon, W., 194
Crosby, A. W. Jr., 192
Cross, W. R., 84, 127, 200
Cuba, 19
Cubans, 28, 29, 163
Cuff, D., 204
Cultural convergence, 85–88,
    110
Cultural divergence, 85–88
Cultural geography, 187–88
    domain of, 74–76
    status, 144
Cultural processes, 5–9
    population mobility as a factor
        in, 77–79
Cultural systems, 3, 40–41
    change in, 53–58
    genetic drift of, 6
    local evolution, 6, 67–68, 111
    origins, 35, 38–39, 45
    relationships with habitat,
        5–6
    socioeconomic components
        and correlates, 8, 55, 72,
        103, 105, 116–17

Cultural systems (*cont.*)
  spatial interactions among, 8, 76, 86
  structure, 71–74, 73 (*diagram*), 87–88
Cultural transfers:
  aboriginal, 14–17
  African, 18–20
  Continental European, 20–22
Culture:
  artifactual aspects, 73–74
  definition, 68–74
  mentifactual aspects, 73–74
  sociofactual aspects, 73–74
  transnationalization of, 156–60
Culture areas, 76–77, 109–39, 118–19 (*map*)
  classification, 110–13
  definition, 112–13
  internal structure, 113–16
  spurious, 112
  synthetic, 181–82
  traditional, 110–11, 177–79
  voluntary, 134–39, 181
Culture core, 39–40
Culture hearths, 83, 89, 117
Culture shock, 24, 71
Cummings, R. O., 201
Cunliffe, M., 147
Curti, M., 160
Curtis, J. R., 169
Cussler, M., 201
Czechs, 163

D

Dallas, Tex., 166
Deffontaines, P., 95
DeGive, M. L., 201
Delaware, 126
Delaware Valley, 20, 83, 125
Demography, 11, 103, 161, 165, 209
Denevan, W. M., 195
Denver, Colo., 133
Depression, 1930s, 144
De Rochemont, R., 202

Deskins, D. R., Jr., 28
Detroit, Mich., 21, 28, 77, 78, 133, 176
Deutsch, K. W., 35
Dicken, S. N., 4, 187
Dickens, C., 169
Diet, 103, 105, 159, 162, 200–201
Diffusion of innovations, 79–85, 156
Dinnerstein, L., 195
Disciples of Christ, 94, 99
Doeppers, D. T., 28, 133
Donley, M. W., 208
Dorson, R. M., 107, 202
Doukhobors, 11
Driver, H. E., 46, 195
Dunbar, G. S., 83
Duncan, J. S., Jr., 169, 196
Duncan, O. D., 69
Dutch, 20, 23, 24, 25, 28, 125, 127
Dutch Reformed Church, 98
DuWors, R. E., 204

E

Eagle, 151, 152
Eastern Shore, 180
Economic conditions, 144
Economic patterns, 50
Edge Cities, 166, 168, 181
Education, 51–52
Educational subregions, 136
Eire, *see* Ireland
Ekirch, A. E., 194
Elazar, D. J., 201
Elderly population, 161
Eldridge, H. T., 87, 152
Electrification, rural, 165
England, attitudes toward, 209
English First, 175
Environment:
  attitudes toward, 193–94
  impact upon culture, 6–7, 29, 60–61, 127
  man's impact upon, 16, 60–61

Environmental Movement, 162
Equine regions, 137
Erikson, J. L., 157
Ethnic cuisine, 158, 202
Ethnic groups, 18–28, 38–39,
    195–96
  spatial distribution, 26, 28–29,
    30–31 (*map*), 32–33, 127
Ethnicity, 149
Ethnic Revival, 173–77
Evangelical and Reformed
    Church, 98
Evangelical United Brethren
    Church, 98
Evans, E. E., 11
Exurbanization, 165–66

F

Falklands, 9
Family structure, 19
Fashion, 82
Faulkner, W., 169, 209
Fellows, D. K., 28
Fences, 101
Ferguson, C. A., 199
Ferris, W., 206
Festivals, 175–76
Field, J. A., Jr., 157
Field boundaries, 102, 197
Fighiera, G. C., 157
Filipinos, 163
Films, 148, 153, 170
Finns, 20, 21, 29, 125
Firearms, 53, 148
First Effective Settlement, Doc-
    trine of, 13–14, 20, 23, 34,
    81, 192
Fischer, D. H., 192
Fischer, E., 209
Fishman, J. A., 199
Fisk, M., 52
Flag, American, 95, 147, 151,
    152, 160
Fleming, D. H., 163
Flemings, 20, 25
Fletcher, S. W., 204

Florida, 21, 24, 29, 32, 93, 109,
    114, 125, 135, 137, 175
Folklore, 106–7, 202–3
Foodways, *see* Diet
Foster, G.M., 6
Francaviglia, R. V., 63, 131, 208
France, 11, 12, 175, 182
Franklin, B., 128
Fraternal orders, 152
Fredericksburg, Tex., 174
Freidel, F. B., 186
French-Americans, 12, 20, 24,
    25, 28, 31 (*map*)
Friis, H. R., 192
Frontier, settlement of, 33–34,
    80–81, 81 (*map*), 82–83,
    129, 144–45
Frontier Myth, 42
Furnas, J. C., 189
Fussell, P., 189

G

Galtung, J., 157
Gans, H. J., 164
Gardens, 92
Garreau, J., 166, 179, 196, 203
Garrett, W. E., 188
Gastil, R. D., 203
Gaustad, E. S., 94, 200
Gay and Lesbian topics, 161
Gay neighborhoods, 181
Genealogy, 177
Gentilcore, R. L., 48
Gentrifying neighborhoods, 181
Georgia, 12, 15, 19, 24, 25, 90
Germans, 25, 26, 29, 30 (*map*),
    146 (*see also* Pennsylvania
    Germans)
Germany, 11, 145
Gettysburg, 160
Giddens, A., 183
Gillin, J., 191
Glass, J. W., 100, 101, 126,
    204–5
Glassie, H., 89, 100, 120, 196,
    202–3

Glazer, N., 32
Glenn, N. D., 87
Gold Rush, California, 131
Goldberg, M., 148, 191
Golf, 158
Good, J. K., 180
Gordon, M. M., 32, 195
Gottmann, J., 37, 60, 62, 205
Gowans, A., 83, 196
Graffiti, 171
Grand Army of the Republic, 146
Grand Strand, S.C., 180
Grasslands, 17
Gravestones, 83, 177, 210
Great Britain, 12, 23, 94
Greater European Cultural Realm, 5, 38, 39
Great Lakes, 20, 21
Great Plains, 61, 208
Greeks, 25, 29, 164
Greeley, A. M., 195
Green, R. W., 43
Greenwich Village, 138
Groves, P. A., 190
Gullah dialect, 19
Gun Lobby, 148
Gypsies, 39

H

Hägerstrand, T., 79
Haitians, 163
Hale, D., 132
Hale, R. F., 180
Hall, E. T., 54, 209
Hallowell, A. I., 16, 195
Hampton Roads, 75
Hamtramck, Mich., 11
Handicrafts, 203
Handlin, O., 22, 37, 195
Hankey, C.T., 129
Hansen, M. L., 22
Hanson, S., 162
Harding, C., 176
Hardy, T., 169
Harries, K. D., 148

Harris, C. D., 186
Harris, R. C., 6, 48, 188
Hart, J. F., 77, 101, 102, 122, 197, 205
Hartshorne, R., 28
Hawaii, 121, 164
Hearst, W. R., 42
Heath, S. B., 199
Hecht, M. E., 169
Hecock, R. D., 169
Hedrick, U. P., 17
Helgren, D. M., 169
Heliotropic regions, 137
Herbers, J., 166
Hero worship, 150, 153, 160
Herskovits, M., 18, 195
Highway strips, 173
Hilliard, S., 105, 202
Hillside letters, 170
Hispanic-Americans, *see* Spanish-Americans
Historical re-enactments, 177
Historic preservation, 170
Historic Revival, 173–77
Historic villages, 182
Hitchens, C., 209
Hofstadter, R., 33
Holbrook, S. H., 84
Holidays, 150, 152, 153, 160, 176
Holloway, M., 63, 210
Homicide, 103–4, 104 (*map*)
Honigman, J. J., 38
Hospitals, 52
Hostetler, J. A., 205
Hotels, 91
House morphology, 88–94
House-types, 17, 196–98
Houston, Tex., 115
Hsu, F. L. K., 43–44, 93, 191
Huber, R. M., 40
Hudson, J. C., 207
Hudson Valley, 20, 121, 127
Hugill, P. J., 169
Huguenots, 12, 20
Human ecology, 69 (*see also* Environment)
Hungarians, 163
Huth, H., 194

Hutton, D. G., 207
Huxley, J. S., 73

I

Icelanders, 29
Idaho, 132, 137
Illinois, 114, 207
Immigration, 22–28, 78, 131,
  162–63, by region and pe-
  riod, 24–27, 27 (*table*)
Indiana, 114
Indian old-fields, 16
Individualism, 41–53, 93
Inkeles, A., 157
Inland Empire, 132–33
Innovations, *see* Diffusion of
  innovations
Interstate Highway System, 156,
  165
Iowa, 114
Iranians, 163
Iraquis, 163
Ireland, 9, 12, 150, 175
Irish, 23, 25–28, 31 (*map*), 163
Iroquois, 17, 117
"The Islands", 179
Israel, 146, 150, 182
Italians, 25, 26, 39, 31 (*map*)
Italy, 12, 182

J

Jackson, J. B., 57, 103, 167–68,
  173, 187, 197
Jackson, K. T., 192
Jackson, P., 188
Jakle, J. A., 28, 133
Japan, 150
Japanese, 29, 30 (*map*), 78, 163,
  175
Jazz, 19, 53, 158
Jefferson, T., 47, 151, 160
Jehovah's Witnesses, 159
Jews, 11, 25, 28, 29, 31 (*map*),
  78, 95, 98, 146, 163, 175

Johnson, H. B., 96, 192
Jones, G., 14
Jones, H. M., 191
Jones, M. A., 22
Jordan, T. G., 21, 22, 28, 123,
  124, 179, 188, 193, 195, 197,
  205, 206

K

Kalamazoo, Mich., 133
Kalm, P., 189
Kammen, M., 210
Kansas, 26, 128, 134
Kariel, H. G., 103
Karn, E. D., 193
Kaskaskia, Ill., 21
Kasperson, R. E., 107
Kaups, M., 21, 193
Kentucky, 85
Kentucky Bluegrass, 124, 137
Key, V. O., Jr., 201
Kielbowicz, R. B., 153
Kimball, S. T., 189
Kishicoquillas Valley, 133
Kluckhohn, C., 69–70
Kniffen, F., 89, 101, 123, 197
Knopp, L., Jr., 161
Kohn, H., 35, 210
Kolodny, A., 194
Koreans, 163
Korean War, 147
Kouwenhoven, J. A., 85, 103,
  203
Kramer, F. R., 49, 126, 207
Kroeber, A. L., 3, 15, 68, 69–70,
  82, 117, 188, 195
Kroeber, C. B., 10
Kurath, H., 105, 120, 121, 133,
  199, 204

L

Labovitz, S., 87
Lafayette, Marquis de, 151
Laird, C., 199

Lake Tahoe, 138
Lamme, A. J., 2nd, 180
Lancaster County, Penna., 176
Landes, D. S., 69
Landgraf, J. L., 130, 208
Landscape, 102–3, 173
  attitudes toward, 169–73
  built, 196–99
Language, 14, 16, 32, 40, 71,
    105, 107, 159, 175, 199–200
Las Vegas, Nev., 138
Latin America, 9, 23, 25, 26, 96,
    163
Lauria, M., 161
Lawns, 92
Lazarus, J., 78
Lee, D. R., 162
Leighly, J., 210
Lemon, J. T., 21, 205
Lerner, M., 190
Levenstein, H. A., 202
Lewis, G. M., 208
Lewis, P. F., 88, 89, 166, 197,
    206
Ley, D., 188
Libraries, 48, 176, 211
Lieberson, S., 195
Liebs, C. H., 197
Lilliard, R. C., 203
Lincoln, A., 160
Linden, F., 106
Linguistic areas, 105
Lipset, S. M., 33
Literature and place, 144, 169
Lithuanians, 164
Little Egypt, 180
Livestock systems, 22, 129
Log buildings, 21, 90
Long, L. H., 56
Long Island, 121
Lonsdale, R. E., 165
Los Angeles, Calif., 166, 167,
    207, 208
Louder, D. R., 189
Louisiana, 20, 21–22, 123
Lowenthal, D., 37, 54, 55, 103,
    137, 177, 194, 197
Lowry, M., 169
Loy, W. G., 208

Luce, H., 145
Ludwig, A. I., 83, 210
Lukacs, J., 145
Lukermann, F. E., 152
Lundberg, F., 46, 190
Lutherans, 98
Lutwack, L., 169

M

MacKenzie, S., 162
Maine, 15, 28, 113, 121, 179
Maize, 16, 17
Mangus, A. R., 120, 203
Manitoba, 128
Manufacturing, 165
Manufacturing Belt, 117
Maoism, 95
Mardi Gras, 176
Markusen, A., 146, 203
Marschner, F. J., 48, 102, 193
Marsh, J. S., 169
Martis, K. C., 201
Marx, L., 49, 64, 166, 194
Marxism-Leninism, 95
Maryland, 21, 32, 127, 137
Massachusetts, 12, 204
Massey, D., 174
Mather, C., 198
Mattson, M., 195
Mayer, H. M., 207
Mazey, M. E., 162
McAlester, V. and L., 196
McCready, W. C., 195
McDavid, R. I., Jr., 199
McGiffert, M., 38, 191
McHarg, I. L., 102
McManis, D. R., 14, 193
McNee, R., 161
McWilliams, C., 132
Mead, R. O., 8, 210
Mechanistic world vision, 59–
    61, 93
Meetings, international, 157
Megalopolis, 117, 137
Meinig, D. W., 10, 22, 63, 114–
    15, 120, 124, 130, 131, 133,
    179, 190, 197, 203, 206, 208

Meinig Model, 114–16, 123
Melting Pot, 32
Melville, H., 37
Mencken, H. L., 191, 199
Mennonites, 12, 133
Mercer, J., 148, 191
Meredith, R., 37, 210
Merrens, H. R., 47, 206
Merritt, R. L., 35, 201
Mesick, J. L., 36
Messianic perfectionism, 61–64
Methodists, 98
Metropolex, Tex., 180
Metropolitan government, 46
Mex-America, 179
Mexicans, 21, 22, 24, 163, 171
Mexico, 179
Meyrowitz, J., 210
Miami, Fla., 28, 46, 179
Michigan, 28, 29, 91
Michilimackinac, Mich., 21
Middle West, 21, 26, 61, 85, 99,
    117, 127, 128–29, 131, 179,
    180, 181, 207
Midland, 83, 84, 100, 116, 117,
    125–28, 204–5
Migration:
  chain, 12, 29
  internal, 32, 56, 77
  selective, 5, 11–13, 78–79
Mikesell, M. W., 3, 10, 188
Militarism, 146–48
Military personnel, overseas,
    157
Military subregions, 135–36
Miller, E. J. W., 108, 124, 203
Miller, P., 37, 54, 191, 210
Miller, T. R., 188
Missions, church, 24, 32, 63–64,
    96, 159–60
Miss Liberty, 151
Missouri, 114
Mitchell, R. D., 98, 190, 203,
    206
Mobility, 53–58, 91
Modernization, 41, 209
  spatial zonation of, 26
Mondale, C., 203
Monk, J., 162

Montane Regions, 137
Monuments, 144, 151, 152
Moravians, 12
Mormon Culture Area, 114, 130,
    131, 155, 177–78 (*map*)
Mormons, 29, 32, 47, 63, 86, 94,
    99, 111, 131, 159, 208
Morrill, R., 79
Moynihan, D. P., 32
Muller, E. K., 204
Multinational corporation, 157–
    58
Mumford, L., 59, 197, 210
Munsterberg, H., 62, 121, 191
Murals, 171
Museums, 172, 176
Music, 150, 167, 177, 209
Myrdal, G., 77

N

Nairn, I., 102, 194
Names, personal, 87, 150, 177,
    200
Nash, R., 194
Nathan, G. J., 191
Nation, defined, 149
National cemeteries, 160
National character, 36–38, 191–
    92
Nationalism, 149–53, 210
National parks, 151, 153, 160
Nation-state, 38, 70–71, 149–50
Native Americans, *see* Aborigi-
    nal population
Navaho, 17, 175
Negroes, *see* Afro-Americans
Nelson, H. J., 102
Nelson, L., 131
Neo-British cultural community,
    10
Neo-European lands, 9
Nevada, 138
New Brunswick, 24
New England, 6, 14, 24, 26, 43,
    47, 77, 84, 85, 98, 99, 100,
    103, 116, 117, 120–22, 131,
    174, 176, 204

"New England Extended", 84, 121, 127
New France, 5
New Hampshire, 121
New Jersey, 20, 121, 127
New Mexico, 22, 100, 115, 175, 176
New Orleans, La., 83, 133, 138, 206
New Spain, 5, 15, 24, 25, 35
New York City, 25, 28, 54, 85, 90, 133, 138, 166, 167
New York (State), 24, 84, 107, 117, 121, 127, 131, 179; 200, 205
New Zealand, 9
Noble, A. G., 198
Nordhoff, C., 63
North Carolina, 14, 20, 47, 122, 127, 206
North Dakota, 114
Northwest Ordinance, 48
Nostrand, R. L., 22, 208
Nova Scotia, 121, 123

O

O'Hara, J., 128
Ohio, 102, 114, 127
Oklahoma, 114, 117, 124, 130
Oldakowski, R. K., 180
Old World, cultural legacy of, 10–14
Olmsted, F. L., 206
Olson, S. H., 205
Olympic Games, 152
Ontario, 29, 121, 128
Oregon, 121, 132, 208
Ouachitas, 124
Ozarks, 21, 108, 124

P

Packard, V., 191
Parsons, J. J., 170
Pastoralism, 64

Patriotism, 149
Pattison, W. D., 102, 198
Patton, C. P., 208
Paullin, C. D., 12, 13, 188
Peace Corps, 63, 160
Penn, W., 15, 47, 125
Pennsylvania, 11, 20, 25, 26, 28, 32, 63, 84, 126–27, 128, 174, 176, 204, 205
Pennsylvania Germans, 20, 21, 99, 112, 126
Perception:
  landscape, 103
  place, 160–62
Periodicals:
  ethnic, 175
  regional, 174
Persian Gulf War, 194
Philadelphia, Penna., 25, 47, 60, 125, 126, 133, 138, 204
Philanthropy, 52, 63, 160
Phillips, K. P., 107, 201
Photography, 170
Pierson, G., 191
Pillsbury, R., 102, 126, 202
Pitts, F. R., 4, 187, 194
Place-names, 16, 63, 105, 107, 150, 152, 167, 199, 200
Planning, 45–46, 170, 175
Plants, domesticated, 16–17
Pleasuring Places, 136–38
Poland, 150
Poles, 25, 28, 175
Political characteristics, 44–47, 49, 107, 152, 161, 200
Polynesians, 164
Population quality, 13
Porches, 92
Porter, P. W., 152
Portuguese, 164
Postman, N., 210
Potter, D. M., 192
Powell, L. C., 37
Powell, S. C., 6, 204
Prairie du Chien, Wisc., 21
Pre-Columbian cultural contacts, 14
Pred, A., 79, 81
Presbyterians, 12, 98

Presidency, 151
Price, E. T., 48, 84, 102, 117, 198
Prince, H. C., 187
Princeton, N. J., 181
Progress, idea of, 173
Prostitution, 161–62
Protestant Ethic, 42–43
Psychological factors, 13, 38
Public opinion polls, 87
Pudup, M. B., 174
Puerto Ricans, 24, 28, 39, 163, 171
Puget Sound Region, 132–33

Q

Quakers, 12, 98, 126, 204
Quebec, 11, 24, 48, 95, 100, 174
Québecois, 179
Quigley, C., 69

R

Racial attitudes, 162
Racial groups, *see* Ethnic groups
Racial hybridization, 15, 177
Radio, 153, 154, 175, 177
Rainey, R., 147
Raitz, K. B., 180, 206
Raleigh Colony, 14
Rapoport, A., 88, 198
Raup, H. F., 21
Rectangular land survey, 47, 60, 102
Reed, C. E., 105, 200
Reed, H. H., 198
Reed, J. S., 180, 206
Regional Revival, 173–77
Regions, 109–39, 203–9, 174, 177–82
    clusters, 183
    social-psychological, 180–81
    synthetic, 181–82
    vernacular, 179–80

voluntary, 111–12, 134–39, 181
Reilly, C. J., 14
Reiners, D. M., 195
Religion, 12–13, 17, 48, 51–52, 94–100, 200
Religious regions, 97–99, 97 (*map*)
Remote sensing, 154
Reno, Nev., 138
Reps, J. W., 102, 198, 208
Research Triangle, N. C., 180, 181
Retirement Regions, 137
Rhineland, 11, 23
Rhodesia, 9
Richardson, J., 82
Rifkind, C., 198
Riley, R., 198
Rinschede, G., 200
Rio Grande Region, 130–31, 135
Roepke, H. G., 165
Romanians, 164
Rooney, J. F., Jr., 189, 202
Root, W., 202
Rosenberry, L. K. M., 84, 121, 204
Rostlund, E., 16
Rowntree, L., 188
Rural Free Delivery, 153, 165
Russians, 20, 25

S

St. Louis, Mo., 133
Ste. Genevieve, Mo., 21
Sale, K., 194, 210–11
Sale, R. D., 82, 193
Salt Lake City, Utah, 95
Salvadorans, 163
Sandberg, L. A., 169
Sanders, W. T., 68
Sanford, C. L., 62, 64, 194
San Francisco, Calif., 90, 100, 131, 133, 138, 166
Sauer, C. O., 14, 79, 84, 210
Savannah, Ga., 83, 133

Scandinavians, 14, 23, 26, 28, 30 (*map*), 32, 98
Schein, R. H., 152
Schlereth, T. J., 198
Schmitt, P. J., 64
Schools, 48–49
Schwartzberg, J. E., 112
Scientific thought, history of, 69
Scotch-Irish, 11, 20, 23, 25, 29
Scots, 20
Sea Islands, 123
Sealock, R. B., 200
Seaside, Fla., 181
Seattle, Wash., 138
Seely, P. A., 200
Séguin, R. L., 100–101
Settlement patterns, 47–49, 58, 75, 95, 100–102, 164–69
Sexual mores, 161–62
Seyler, H. L., 165
Shemanski, F., 176
Sherman, W. C., 196
Shoemaker, A. L., 101
Shopping centers, 58, 168, 171–72
Shortridge, B. G., 189
Shortridge, J. R., 179, 200, 207
Shurtleff, H. R., 21, 84
Siegel, B. J., 38
Siegfried, A., 37, 137, 190
Sierras, 137
Signs, 144, 154, 170–71
    welcoming, 170
Silicon Valley, 180
Silk, J., 169
Slater, S., 77
Slave Trade, 12, 18, 25
Slavic population, 31 (*map*), 78
Sloane, E., 100, 108, 198
Smith, A., 71
Smith, A. D., 164
Smith, E. C., 200
Smith, H. N., 64, 194
Sopher, D. E., 63–64, 95, 99, 200
Sorre, M., 103
South, 32, 61, 99, 100, 109, 117, 122–25, 180, 181, 201, 202, 205–7
    subregions, 123–25

South Africa, 9, 148
South Carolina, 20
Southern California, 7, 29, 102, 109, 132, 135, 208
Southwest, 20, 22, 24, 29, 58, 99, 113, 117, 130–31, 179, 180, 208
Space, perception of, 54–55
Spanish-Americans, 20, 29, 31 (*map*), 130–31, 162, 164, 171
Spencer, J. E., 24, 79, 131, 188
Sport, 144, 159, 202
Stanford, Calif., 181
State, defined, 149
State College, Penna., 181
Statism, 149
Stein, M. R., 37
Steinberg, S., 164
Steiner, M., 203
Steward, J., 39
Stewart, G. R., 43, 105, 190, 200, 211
Stewart, W. A., 19
Stilgoe, J. R., 165, 198
Strauss, A. L., 49
Struik, O. J., 43
Sturtevant, W. C., 196
Success Ethic, 43
Sun Belt, 180
Susquehanna Valley, 83, 125
Suttles, G.D., 28–29, 133, 196
Swain, H., 102, 198
Swedes, 20, 24, 125, 146
Swiss, 20, 25, 146
Symanski, R., 162
Synthetic regions, 181–82

T

Taeuber, C., and I. B., 22, 56
Taeuber, K. E. and A. F., 77
Tager, J., 204
Taylor, G. R., 33
Taylor, J. G., 202
Television, 148–49, 153–54, 159
Tennessee, 114

Texas, 22, 24, 28, 32, 58, 60, 110, 114, 115, 130, 165, 175, 205, 206
Texas Culture Area, 114–16, 116 (*map*), 123, 125 (*map*)
Thai, 163
Theme parks, 171–72, 182
Thernstrom, S., 196
Thomas, D. S., 78
Thomas, W. L., Jr., 4, 79, 132, 188, 208
Thompson, J. H., 205
Thoreau, H., 36, 42
Thrift, N., 174
Thrower, N. J. W., 48, 102, 193
Time, perception of, 54–55
Tobacco, 17, 61, 80
Toffler, A., 53, 112
Toqueville, A. de, 37, 39–40, 120–21, 192
Tourism, 156, 172
Trachtenberg, A., 211
Tracy, S. J., 22
Trailers, 91, 156
Trans-Atlantic migration: cultural effects of, 5
Transnationalization of culture, 156–60, 184–85
Transportation, 50–51, 57–58, 155–56
Trépanier, C., 206
Trewartha, G. T., 47, 102, 198
Tuan, Yi-Fu, 92
Tunnard, C., 103, 198
"Turnaround", metro-nonmetro, 166
Turner, E. J., 195
Turner, F. J., 193
Turner, L. D., 19
Turner Thesis, 33–35

U

Ujifusa, G., 200
Ukrainians, 164
Ulack, R., 180, 206
Ullman, E. L., 211
Uncle Sam, 151

Unitarians, 98
United States: atlases, 188–89
significance of, 3–5
(*see also specific topics*)
University of Chicago, 182
Urban morphology, 102
Utah, 32, 103, 114, 115
Utilitarianism, 59–60
Utopianism, 62–63, 138–39, 152, 210

V

Vail, W. G., 36
Vance, R. B., 103, 122, 206
Velikonja, J., 29
Vermont, 28, 114, 121, 127, 179
Vernacular regions, 179–80
Veterans associations, 153
Veysey, L., 211
Vietnamese, 163
Vietnam War, 147
Vincennes, Ind., 21, 48
Violence, 146–49
Virginia, 12, 47, 48, 83, 114, 127, 137, 205, 206
Virgin Islanders, 24, 39
Volga Germans, 11
Voluntary associations, 52, 107
Voluntary regions, 111–12, 134–39, 181

W

Wacker, P. O., 134
Wade, R. C., 207
Wagner, P. L., 3, 4, 188
Wallach, B., 194
Ward, D., 193
Warner, W. L., 190
Washington, D. C., 137, 138, 166, 179
Washington, G., 160
Washington (State), 121, 132
Wasserman, P., 176

Waters, M. C., 195
Watson, J. W., 62, 190
Webb, W. P., 22, 37, 208
Webber, M. M., 112
Weber, M., 212
Weightman, B. A., 161
Weiss, M. J., 183, 190
Welsh, 20, 23, 25, 126, 164
Wertenbaker, T. J., 37, 126, 205
West, 99, 129–33, 179, 207–8
West, R. C., 105
West Africa, 11, 18
West Virginia, 28, 114, 127
Wheeler, J. O., 28
Wheeler, T. C., 190–91
White, L. A., 68
White, M., and L., 49
White, S. E., 166
Whitehill, W. M., 204
Whitney, E., 59
Wilkie, R. W., 204
Willamette Valley, 132–33
Williams, C. H., 164
Williams, M., 193
Williamsburg, Va., 54, 82, 182
Wilson, C. R., 206
Wilson, L. R., 211
Wisconsin, 26
Witney, D., 196

Women's Movement, 161–62
Wood, G. R., 123
Wood, J. S., 166–67, 204
Woodruff, W., 211
Wright, A. T., 37
Wright, J. K., 12, 13, 37, 188, 211
Wyman, W. D., 10
Wyoming, 85

Y

Yard decorations, 171
Yellow ribbons, 147
Yoder, D., 105
Young, W. J., 204

Z

Zelinsky, W., 28, 57, 77, 87, 89, 90, 94, 97, 99, 101, 105, 107, 123, 124, 127, 148, 149, 152, 154, 158, 162, 170, 180, 189, 191, 199, 200, 202, 203, 204, 205
Zimmerman, C. C., 120, 204